Engineering Graphics Desk Book

Thomas T. Samaras

Engineering Graphics Desk Book

Prentice-Hall, Inc. Englewood Cliffs, N.J.

Prentice-Hall International, Inc. *London*
Prentice-Hall of Australia, Pty. Ltd., *Sydney*
Prentice-Hall of Canada, Ltd., *Toronto*
Prentice-Hall of India Private Ltd., *New Delhi*
Prentice-Hall of Japan, Inc., *Tokyo*

PRENTICE-HALL, INC.
Englewood Cliffs, N.J.

Library of Congress Cataloging in Publication Data

Samaras, Thomas T
Engineering graphic desk book.

Includes bibliographical references.
1. Engineering design--Management. 2. Engineering graphics--Management. I. Title.
TA174.S25 658'.91'6042 75-17554
ISBN 0-13-277855-6

Printed in the United States of America

TO MY WIFE,
DR. MARY STENNING SAMARAS

TO MY PARENTS,
THEODORE AND GARYFALIA SAMARAS

ACKNOWLEDGEMENTS

Special thanks are given to James Samaras, Actron Industries, and my friend and associate, Donald Gould, Manager Engineering Services, Astrodata, Inc. for their comments and recommendations on the manuscript.

What Will This Book Do For You

Millions of engineering drawings are produced annually in the U.S. alone. And while books have been published on the mechanical and mathematical techniques for converting ideas and technical information into graphical or pictorial forms, no book until now has been written that covers the entire spectrum of activities related to the non-drafting aspects of engineering data. Although neglected, this area of engineering graphics is essential for the efficient preparation, use, and protection of very costly documents.

Because of the growing importance of new techniques and good management to the preparation of engineering drawings and related data, this book was written to provide industry with a reference book on these important cost and performance areas. It is especially dedicated to technical people in small and medium size companies who do not have the facilities and specialists available for becoming aware of the many advances in technology and techniques that have taken place during the past decade.

This book gives 10 areas for profit and quality improvement through the application of systematic techniques in the design and drafting operation. It will also help you to avoid serious problems that can wipe out budgets, create dissatisfied customers, and develop internal difficulties leading to lower morale and efficiency. Any organization, whether it has a 1-man or 200-man design and drafting staff can benefit by applying the techniques and information presented in this book to its daily operations.

Why do you need this book? What will it do for you? Answers to these questions are provided in the following checklist:

- This book tells you how to take a systems or total viewpoint towards the document preparation task. This is done by examining the various life cycle operations involved and by seeing how their requirements affect drawing preparation and usage. *See Chapter 1.*
- This book tells you how the design and drafting operation is organized. It also gives staff requirements, control needs, and techniques for improving efficiency. Automated drafting systems are discussed. *See Chapter 2.*
- This book tells you how to protect your engineering data from loss, damage, and unauthorized changes that can lead to product failure or unneeded expense. *See Chapter 3.*

- This book tells you about the various kinds of drawings, specifications, and procedures used by many companies. It provides sample drawing formats to illustrate their requirements. *See Chapter 4.*
- This book tells you how to identify your documents and the objects built from them. It describes different types of number systems and their value. Also guidelines are presented for re-identifying parts or items so that you don't deliver a part to your customer that is dangerous or fails to meet his needs. *See Chapter 5.*
- This book gives you information on how to control data and changes to minimize compromising the design, creating errors, and problems in the shop. Various forms used to document and control changes are described. *See Chapter 6.*
- This book gives you a review of various types of records, reports, and microforms used in industry. Status cards, log books, and change status summaries are discussed. *See Chapter 7.*
- This book gives you a review of the various types of techniques available for reproducing drawings. The reproduction cycle is examined and methods for improving reproduction operations described. Methods for reducing costs are presented. *See Chapter 8.*
- This book gives you a review of the cost of design and drafting and presents different methods for estimating costs. Sample costs for different types of documents are given. *See Chapter 9.*
- This book covers management control systems, schedules, and computer reports. Both manual and automated control systems are reviewed. *See Chapter 10.*

HOW THIS BOOK IS ORGANIZED

Each of the 10 chapters covers a key area of the design and drafting (D&D) operation and blends general descriptive data with the benefits offered by various D&D techniques. Numerous real-life examples are presented to illustrate the value of various techniques in reducing costs or improving product quality.

The book is oriented towards all sizes of companies. Small, medium, and large companies have the same goals, needs, and problems. Only their complexity and degree vary with size.

Numerous illustrations, sample forms, checklists, and tables are presented to help in your day-to-day operations and in applying new techniques to your organization when the need is evident.

A detailed index is provided at the end of the book to simplify finding a specific area of interest or new technique. Sources for additional information on D&D activities are given in Chapter 2 and in the Appendix.

HOW TO USE THIS BOOK

ENGINEERING GRAPHICS DESK BOOK is a read-through and reference volume. It can be used as follows:

1. If you want to reduce the cost of preparing engineering drawings, this book will provide you with various techniques to evaluate in terms of your own operation. Techniques such as composition drafting, ink drafting, automated graphics, word processing, job shoppers, and reproduction equipment are discussed.

2. If you have a problem and need a method to solve it, look at the table of contents for the general subject area. Then review the material to see what techniques have been used by others in industry. From this information and an objective evaluation of your needs, you can design a system to meet your requirements.

3. If you need to broaden your understanding of the D&D operation because you are new to the field or have had a narrow range of experience, read the entire book, studying those topics that have most immediate application in your present job.

4. If you are interested in being a supervisor or manager of a D&D operation, study this book completely. It will provide you with the basic framework to understand the overall requirements of the D&D operation so that you can better manage its activities.

5. If you are an engineer or manager who has to work with D&D, this book will enable you to understand the needs and problems of this department and how to improve your ability to communicate with its people.

6. If you need information on D&D operations from time to time, use this book as a handy reference book to provide you with quick information. For example, pay scales, costs of drafting, sample forms and data, and trade journals are some of the topics that are immediately available to you when needed. Use the table of contents and index to find information needed.

Most design and drafting departments are not operating at optimum efficiency. A prime reason for this condition is a nonsystems approach to the preparation and use of engineering data. Other reasons include lack of awareness of cost reduction techniques, poor coordination among the various operations within the company, and inattention to management control systems.

All design and drafting departments can be less costly and more productive. This book offers a variety of proven techniques for achieving these critically needed qualities.

Thomas T. Samaras

Contents

Engineering Graphics Desk Book

1

1.0 Engineering Graphics: Special Tools for Improving Performance and Reducing Costs

Engineering graphics is a discipline that can be divided into various subject matter specialties such as mechanical, electrical, chemical, structural, civil, and architectural. Although these specialties are interconnected by a common core of graphic skills and techniques, each requires unique talents and expertise for efficient conversion of engineering concepts into accurate and clear physical descriptions and drawings. The non-drafting aspects of engineering graphics, however, represent a universal center of skills and methods that transcend the boundaries of these specialized fields. For example, design and drafting (D&D) operations, data control, change control, record keeping, and reproduction of data are activities that are subject to the same principles, regardless of the particular subject matter of the documents produced. Note that the terms "data" and "documents" are used interchangeably throughout this book; however, "engineering data" is restricted to the more narrow category of specifications, drawings, and associated lists. Also the term "product" is used to identify virtually anything that requires drawings to make it. Thus, "product" refers to a mechanical or electrical part, a building, a car, a bridge, or any other man-made item.

A draftsman or designer spends only a fraction of his day in actual drafting and design work. As a result, the non-drafting aspects of engineering graphics are of immense importance to him and to the success of the entire company.

CASE IN POINT: One medium size company examined the utilization of its senior draftsmen and found that they spent only 3 to 4 hours a day on actual drafting work. The rest of the time was spent in collecting information, obtaining checkprints, filling out status cards, and interfacing with supervisors and engineers.

1.1 Non-Drafting Aspects of Engineering Graphics

Some examples of the non-drafting aspects of engineering graphics were already given. However, a more detailed review of these aspects is worthwhile in obtaining a sharper picture of their relations and importance to the D&D operation.

DESIGN AND DRAFTING DEPARTMENT: KEY TO SYSTEMATIC OPERATIONS

The design and drafting department is, of course, the center of interest here because it provides the direction, framework, capital, and facilities for translating ideas and designs into graphical portraits that can be used by others to re-examine their concepts, to build new products, to test and install products, or to understand how products work. While accurate, complete, and rapid production of quality engineering data is the prime effort of this department, many other functions must be performed or monitored to ensure the ultimate utility of the data produced (for example, drawings, parts lists, bills of materials, and wire lists). Therefore, the design and drafting (D&D) department needs to be examined in terms of other functions that are vital to its success from a systems viewpoint. That is, an effective department is one that intermeshes efficiently with associated departments to achieve maximum output and performance for the company as an entity.

CASE IN POINT: One medium-size company found failure of the D&D department to intermesh effectively with the rest of the company expensive and politically untenable. Lack of communication and response to the needs of other departments created conflicts between the D&D manager and other managers. As a result, preparation and alteration of engineering data were not given special attention to minimize delays in purchasing and manufacturing and hardware could not be shipped on schedule. This cut into company sales and

profits and finally top management began looking into the problem. A new manager, sensitive to the needs of others, was hired and improved relations were quickly restored.

Another important aspect to be considered is the ability of the department to minimize its growth by increasing its productivity or efficiency. Therefore, growth by itself should never be considered an absolute measure of success. What really counts is the total output and the ratio of output to input, or efficiency. For this reason, cost reduction and performance improvement are key objectives of the D&D department.

DATA CONTROL: KEY TO DOCUMENT INTEGRITY

The data control department is, or should be, an autonomous body within the company that is responsible for the care, storage, protection, retrieval, and control of released engineering data. The phrase "released engineering data" indicates that engineering and D&D have officially approved the issuance and use of the data for production, procurement, or testing. (See Chapter 3 for further discussion of these terms.)

Without a data control department, the integrity of drawings, specifications, and parts lists cannot be assured. Because it is the keeper of the D&D department's output, it plays a key role in the daily activities of draftsmen and designers. Therefore, the objectives, responsibilities, method of operation, and problems of the data control department must be understood to increase the effectiveness with which documents flow between these two departments and other parts of the company. Failure of D&D personnel to understand the requirements and importance of the data control function can lead to a rough interface that irritates all involved and reduces productive cooperation and communication between the two departments. It can also lead to uncontrolled changes, increased errors, higher costs, lost documents, and production delays.

SYSTEMATIC DOCUMENTATION OF ENGINEERING DATA

Drawings and engineering data are prepared in various types, sizes, and formats. Therefore, engineers, D&D personnel, and production people need to be familiar with the purpose, scope, and requirements of each document for an efficiently operating company. Drawings, for example, have several classifications, e.g., detail, assembly, control, and installation. They are also categorized by uses, e.g., procurement, manufacture, and design evaluation. Besides these classes, engineering drawings can be physically identified as either book or traditional large sheet documents.

Parts lists (bills of materials), data lists, and specifications are different classes of engineering data. These have their own unique formats and requirements and differ considerably from drawings.

The many-sided and complex nature of engineering documentation poses a serious problem to the D&D department and company because confusion, waste, errors, and low production can develop from a random and inconsistent approach to document format, content, and preparation techniques.

IDENTIFICATION OF DOCUMENTS AND PARTS: KEY TO CONFIGURATION CONTROL

Identification of documents and parts is essential to the control and traceability of manufactured items to their parent documents (specifications, drawings, and parts lists). A well-defined system for assigning unique identifiers to each engineering document produced by the D&D department is necessary to assure control of a product's configuration. This system must not only be known by D&D personnel but by other department personnel such as engineering, production, quality assurance, material control, and purchasing.

Although the department fabricating the product is usually responsible for marking it with correct identifiers, D&D personnel must be sure that information is recorded on the drawings to (1) ensure a unique identification of each part or item and (2) define the method of making this identification, e.g., engrave, paint, ink, or rubber stamp.

MANAGING CHANGE TO AVOID TECHNOLOGICAL ANARCHY

Control of alterations to data has become a prime function in industry today because of the complexity of products, short product development lead times, and frequent changes to the product's configuration. A part of this control process involves converting full size vellums into microform for permanent record retention and better control of documents.

Failure to control the identification, preparation, processing, release, and storage of data inevitably leads to loss of knowledge of what the product really is, externally and internally, and what it can do. Sloppy evaluation and control of changes to the product lead to the same results. Also uncontrolled changes can create serious consequences in other subsystems and areas because of the great interdependence and interaction of various items in a product.

CASE IN POINT: One company found that making a seemingly trivial change to a product was more costly than expected. It lost thousands of dollars by substituting a newly

developed part with better performance capabilities for one already ordered. The new order was placed and the old one cancelled. The consequence of this action was a longer delivery period than promised because of manufacturing problems. This held up production 2 weeks. In addition, because the part was newly developed, some defective parts were found after installation and several completed assemblies had to be reworked. As a result, production, installation, and sell-off of the completed product slipped 4 weeks causing additional expenses, late delivery, and customer dissatisfaction.

The preceding example illustrates the need for careful evaluation of the possible ramifications of a change and its value to the company. To minimize problems, make sure that the correct level of management and technical people have the chance to evaluate the impact of the change before it is made. Thus unacceptable perturbations to the schedule and budget are minimized.

RECORDS AND REPORTS PROVIDE VALUABLE INFORMATION

Records and reports are of interest to almost everyone in a responsible position within the company because of their importance to the communication process and necessity for conducting business. A completed record usually indicates that certain procedures are being followed and contains important technical and administrative data. For example, a drawing release record in a data control department indicates the following: (1) the person who submitted an approved drawing vellum to data control for reproduction; (2) who got copies; and (3) that a vellum can be found in a data control file.

Reports, of course, provide interested staff members with information on various actions related to the preparation, revision, and release of data. They also give status, performance, and cost information essential for making management decisions. For large operations, computer information and reporting systems are needed to process and filter extensive data on a bi-weekly or weekly basis and to reduce clerical and accounting labor costs.

Original drawings are often microfilmed as a permanent record before issuing prints to users. This subject is covered in Chapter 7.

Although the preceding comments apply primarily to management, performance evaluation of draftsmen and designers is often strongly influenced by the value of their inputs in preparing reports and records. If the inputs supplied by these personnel are deficient, then management will be making decisions on false information contained in the reports and records. Therefore, effective record-keeping and reporting are valuable characteristics of D&D personnel.

CASE IN POINT: A company found that failure of its draftsmen to properly record status and time data on work cards created confusion and unnecessary costs and interface problems. The cards were kept in a central area near the supervisor so that he could monitor costs and quickly determine the status and location of each drawing. For the system to work, each draftsman had to update the status card for each of his drawings. This required that he add his name and start/stop times to the card when he picked up the drawing and when he completed it. Failure to do this consistently resulted in cards lacking timely information and the supervisor could not provide the current status of a drawing when an engineer, designer, or manager was concerned about its progress. It also took more time for the supervisor to track down the drawing and to determine its location. Cost? An additional administrator was hired to track the status of each document.

DATA REPRODUCTION: KEY TO DISSEMINATING ENGINEERING INFORMATION TO USERS

Reproduction of data is a permanent and costly part of engineering data preparation. Even a 40-man company can require the duplication of thousands of documents per month. While this operation is straightforward and rarely involves major technical difficulties, it can be an expensive and time consuming activity that can affect the efficiency of other groups throughout the company. The availability of timely and readable data is often essential to performing work efficiently in most companies.

COST CONTROL AND SCHEDULES: A MUST FOR SUCCESSFUL PROJECTS

Costs and schedules in today's competitive industrial arena can only be ignored by those bent on failure. D&D and associated costs are often a major part of a company's budget. Projects involving R&D or new product development often set aside more than 30 percent of their total budget for D&D and related engineering data activities. Therefore, accurate cost estimating and effective control are essential operations for successful management and profitability.

Scheduling involves planning the accomplishment of work on a time phase basis. Thus accurate estimates of the number of drawings, quality requirements, hours per drawing, and changes expected are required to determine when a project can be completed. Good scheduling also requires the ability to assess the overall temper of the project and the design problems that will occur. This knowledge allows extra hours and calendar time to be realistically added for projects that will have more than the ordinary share of problems.

1.2 Drawing Life Cycle: A Systems Approach

An important ingredient in the appreciation of the non-drafting aspects of engineering graphics is the understanding of a drawing's life cycle (see Figure 1). Examination of the flow chart shown in this figure immediately reveals that the actual design and drafting operations are only a small part of the entire picture. Thus, the steps taken from the conceptual development of a problem's solution to the final deactivation and destruction of a drawing are strongly interrelated and interdependent, even though design and drafting personnel may not be directly involved.

While the design and drafting aspects are, of course, the keystones in the process, ineffective or deficient completion of the other operations shown can undermine the utility and value of the final drawings made. For example, if a perfect drawing gets to the user late or not at all, its value to the company may be zero. Therefore, the inputs and operations of engineering, data control, production, material control, field service, and management are necessary elements in the life cycle of a drawing.

The key point to remember is that although a particular step may represent a small portion of the labor expended in preparing the drawing, its omission or incorrect implementation can compromise the validity of the document as an unambiguous and complete description of an item's configuration. Consequently, it is important that managers, supervisors, engineers, designers, and draftsmen take a systems or total approach attitude towards engineering drawings and data.

1.3 A Systems Approach Means Better Performance

While the D&D operation is usually viewed as an independent function, the performance sensitivity of so many other departments to D&D requires an overall viewpoint or orientation to ensure maximum benefits for the company. Thus the principal objective of a systems approach to the D&D operation is to achieve company objectives more effectively by concentrating on the effectiveness of the total operation rather than on individual groups within it. That is, it always places top priority on the efficiency of the entire system (D&D

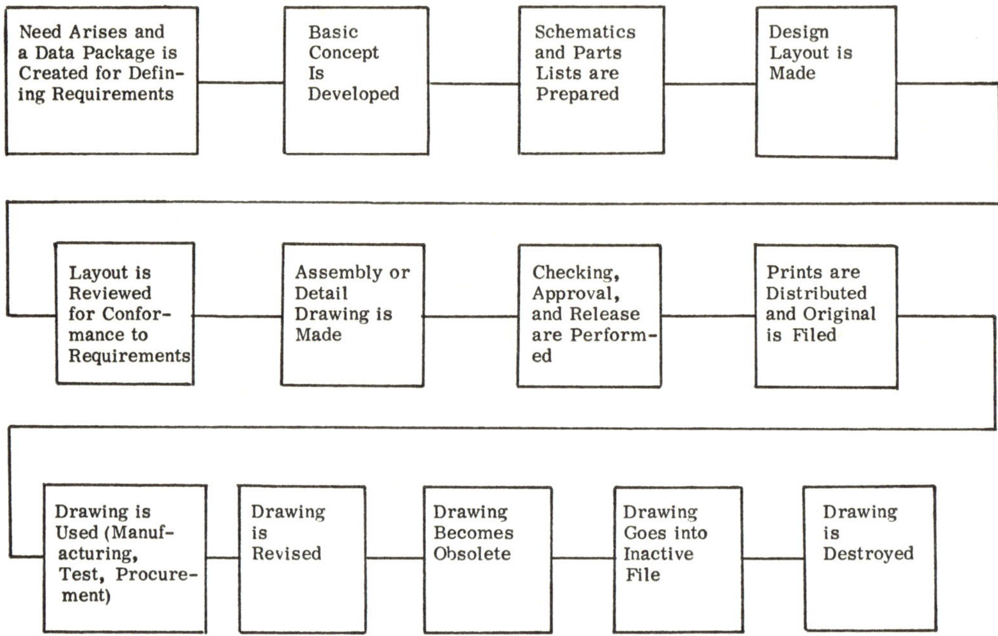

Figure 1 Drawing Life Cycle Flow Chart

plus other departments) even though a particular element or subsystem must perform less efficiently to achieve this end. Secondary objectives for achieving total D&D effectiveness include:

1. More efficient management through increased visibility of activities and more effective control, planning, and organization.
2. Sharper definition of objectives, job descriptions, and operations for clearer and more precise assignments, less wasted effort, and fewer mistakes.
3. Improved integration of different functional elements such as procurement, production, engineering, design and drafting, and inspection.
4. Better information exchange on problems and operations among divergent groups through formal and informal methods. Includes lateral, vertical, and diagonal channels of communication.
5. Faster and more effective response to changing company requirements and conditions.
6. Better control, implementation, and accounting of product design data and changes to its configuration.
7. More accurate prediction and control of costs.

8. More complete and timely supply of documents on system element activities to management.

Figure 2 shows the various departments that are dependent on the design and drafting department for their effective performance.

In implementing a systems approach, the D&D department must be aware of each department's functions, needs for top performance, problems, and effects on other departments. Likewise, each department that interacts with

Figure 2 Functions Dependent on Inputs from Design and Drafting

D&D must understand its needs, functions, problems, and effects on other departments. Although most key people in industry have a fair understanding of what each department in their company does, periodic meetings of senior people can help highlight each department's needs and can also reveal the consequences of past weaknesses in understanding and communication. Then these representatives can go back to their departments and tell the staff what they are not doing right for departments X and Z. Of course, what is being done right should also be covered to reinforce and ensure continuation of good interface activities. Although the need for this communication process seems self-evident, it is often not followed in practice.

An interface chart is shown in Table I to illustrate one way of defining the responsibilities and needs of various functional categories in the company. Note that the definitions will vary from company to company. The interface personnel can also differ, so that the list can be much longer or shorter.

CASE IN POINT: The ignoring of other people's needs to do their jobs is common in industry. In one company, engineering investigated the purchase of a complex piece of equipment so that it could increase its productivity. When a decision to order this equipment was approved, it was found that the purchasing department knew nothing about this type of equipment and was in a difficult position to make the best buy from several available suppliers. As a result, delays in the purchase of the equipment occurred even though top management had given engineering a tight schedule to install the system and get it working.

If engineering had asked a purchasing representative to participate earlier during the evaluation period, purchasing would have been able to respond more quickly and efficiently and engineering's schedule would have been met.

Note that while interface problems are most troublesome in medium and large companies, small companies with only 15 or 30 people also have them, although to a lesser degree.

In taking a systems approach, certain areas are emphasized by management, e.g., the critical thought of department leaders is focused on the interfaces that must be designed and put together to assure smooth and effective document and information flows necessary for each department to optimize its performance. Of course, compromises must be made since conflicting requirements will exist that cause one department to suffer a degradation in performance as a trade-off in the improved operation of one or more other departments. However, in a systems oriented company, this poses no problem when the benefits to be derived by one department at the expense of another clearly benefit the company's overall performance. Of course, the decision to sacrifice a department's performance for the benefit of the company should always be documented in a memo or special report.

When a clear cut choice is not readily apparent for determining that a department's sacrifice in performance will provide an overall benefit to the company, a decision can be made using the following methods:

1. Quantitative or qualitative models of different approaches comparing the benefits and disadvantages of each approach are prepared. Objective merit standards are identified and the results speak for themselves.
2. A review board discusses and analyzes the problem and a majority vote determines method to be used.
3. Top management reviews approach and makes autocratic final decision based on its experience, knowledge, and intuition.

The method selected depends on the management philosophy of the company and its past successes with the technique used.

An important aspect of a systems approach to the D&D operation is the configuration management discipline. This discipline is devoted to assuring that a product meets customer requirements from the viewpoint of form, fit, and function. In addition, a major aim is to know what the configuration of a product is through its entire life cycle, e.g., from its conceptual stage through its production and operational stages. Another equally important objective is to put change control in the hands of the correct technical and management people so that the impact of a change on cost, performance, and schedule can be realistically evaluated. To achieve these objectives, the discipline of configuration management is organized around three basic activities: identification, control, and accounting.

The *identification* function includes assigning unique identifiers to all documents and hardware so that the two can always be interrelated and traced to each other. For example, a particular drawing should be traceable to the item built from it. On the other hand, a completed item should be traceable back to the exact drawing revision used to build it. Of course, the major part of the identification system involves the process of identifying the configuration by graphical and textual means: specifications, drawings, and lists.

Control involves the review, evaluation, and approval of changes to approved documentation defining the product's requirements or configuration. An important aspect of this function is thorough evaluation of the impact of a design change on the end product cost and schedule. It is also vitally concerned with the impact of a change on associated hardware, facilities, and data.

Table I

Table of interface responsibilities and needs*

Interface Personnel	Responsibilities	Needs
Manager/Boss	Gives over-all view of product concept and use. Sets key goals. Provides resources. Approves expenditures and major actions. Solves major problems. Notifies of changes in plans/goals/requirements. Approves major changes in design.	Notification of problems. Status reports New or urgent needs Early notification of expected schedule slippage, failure to meet technical requirements or to stay within budget. Notice of exceptional achievements/progress.
Designer/Draftsman	Prepares schematics, layouts, assembly drawings, parts lists, etc. Identifies documents and assigns part numbers. Follows change control system and procedures. Informs of design simplification and improvement possibilities, weak design areas and potential problems. Meets schedule and budget and reports on status. Meets standards for drawing quality and format. Gets proper approval of design documents before release to manufacturing.	Product requirements. Design goals and potential problems. Special features and potential problems. Schedule and budget. Notification of changes in requirements/design. Special processes or specifications that must be used and test procedure numbers. Parts selection criteria. Manufacturing restraints/considerations. Names of specialists who can help on critical design areas. Test and checkout considerations. Maintenance and repair considerations.
Test Engineer	Designs test setups and gets test equipment and facilities ready. Obtains test fixtures. Schedules testing. Writes test procedures. Detects potential problems. Informs of design changes for simpler testing and checkout. Reports status. Determines environmental test conditions.	Configuration and description of product and usage. Test requirements. Special design features. Mounting orientation, limitations and method. Potential danger areas. Schedule and budget. Notification of changes in requirements/plans.
Quality-Assurance and Reliability Engineers	Ensure that product meets minimum quality and reliability standards/requirements. Notify of design/manufacturing/testing weaknesses and problems. Provide adequate inspection and documentation of incoming parts, material control and hardware. Monitor special processes. Review and approve all documents.	Technical goals and requirements. Budget and schedule. Description of product. Design data. Special features and quality/reliability standards. Applicable specifications. Notification of changes in requirements or plans.
Sales/Marketing Representative	Analyzes market potential and sales strategy. Produces sales literature. Helps set price for product. Informs of problems or needs.	Description of product, especially new or outstanding features. Product specifications. Scheduled availability of product for distribution. Cost of product.

*Samaras, "Man-Man Interface," *Electronic Design*, No. 11, 1971, p. 54.

Interface Personnel	Responsibilities	Needs
Procurement Representative	Purchases parts and materials at minimum cost. Notifies vendors of changes in requirements. Notifies of delays/problems. Ensures timely delivery.	General and special parts requirements. Identification of long-lead-time parts. Names of desired suppliers. Specification control documents. Quality and reliability requirements. Priority of cost, quality and delivery time. Clear-cut notification when substitute parts cannot be used.
Technician	Builds and tests prototype. Evaluates operation and trouble-shoots. Recommends design improvements. Keeps laboratory notebook up to date.	Performance requirements and criteria. Schedule and budget. Special features/problems. Notification of changes in requirements or plans.
Manufacturing Engineer	Provides tooling and fixtures. Acquires special skills. Meets schedule and budget and quality requirements. Informs of methods for improving productibility of hardware. (Productibility refers to accessibility of parts and realistic tolerances and tooling needs).	Clear-cut goals. Accurate, complete and clear production documents. Special product features and needs. Production document release schedule. All materials needed. Identification of special processes or materials to be used—may require special training program. Changes in plans/design. Schedule and budget.
Facilities Engineer	Provides work areas and equipment when needed. Modifies work areas for special needs.	Exact facility and equipment needs. Special environmental conditions. Scheduled need of facilities and equipment. Changes in plans.
Accountant	Issues job numbers. Provides weekly expenditure reports. Processes time cards. Provides cumulative expenditure reports. Notifies management of dangerous budget trends.	Description of work. Budget control and reporting needs. Subtask budgets. Changes in budgets. Expenditure alert levels requiring notifications.
Publications Engineer	Provides documents on time and in budget. Meets document specs. Distributes documents to staff. Informs of problems or needs.	Document requirements, schedule and budget. Description of product and project activities. Special features of product. Changes in design, schedule or other requirements. Specifications to be followed.
Configuration Manager	Sets up system for ensuring that product meets all requirements and for controlling changes to its configuration. Keeps records of changes and approvals. Reports status of changes.	Product description and requirements. Schedule and budget. Special requirements related to form, fit and function. Notification of all proposed changes.

The *accounting* function deals with the proper documentation and filing of all documents related to the configuration of the product. Thus, drawing status lists, engineering orders, change proposals, waivers, deviations, and document status cards are an important part of this function. So is tracking of in-process change documents.

Although D&D personnel do not get involved in all aspects of configuration management activities, they are intimately embroiled in many of its operations and play a vital role in its implementation and successful operation.

1.4 Economic and Efficiency Considerations Depend On Strong Internal Controls

Any company that overemphasizes sales or new product development at the expense of improved internal operations is going to find that the profits from increased sales are neutralized by deteriorating productivity and efficiency. For this reason, the D&D department's performance should never be taken for granted or ignored.

CASE IN POINT: One large company allowed a laissez-faire condition to exist in its D&D department. When problems in costs and delays became too obvious to ignore further, a new manager was assigned to take over the operation.

He soon found that the department had no records of the amount of labor spent on each drawing, the number of times it was revised, or an average cost per drawing. In addition, control of in-work vellums was non-existent and no schedules for completing work were available. As a result only 10 percent of the required drawings were released during the preceding year. The cost impact from a total sales viewpoint was hundreds of thousands of dollars.

While the preceding example came from a fairly large company, small companies can also be almost as inefficient. The following situation occurred in a company with only 20 employees: A technician assembling a new product found several items that would not fit properly unless filing, shimming, or drilling operations were performed. However, instead of notifying the designer that some changes were needed in drawings to avoid repetition of these problems, he simply fixed them and didn't tell anyone. If the company had set up a simple but documented procedure for preparing drawing change requests, this ineffective operation would have been avoided.

Strong and consistent efforts are needed to ensure that the operation's waste, delays, errors, ineffective procedures, and decaying facilities are being combatted. To do this, using a systems approach, requires constant reappraisal and correction of the following areas in terms of changed company objectives, new technological developments, change in work loads, and altered organizational structures:

- Eliminate redundant, obsolete, and useless operating systems and procedures.
- Replace slow and defective equipment.
- Remove obsolete and redundant reports and records.
- Reduce the distribution and number of drawing copies produced.
- Use new techniques, materials, and tools when their results are worth the cost of implementation and use.
- Eliminate unnecessary functions, steps, forms, documents, and quality requirements.

New techniques and equipment for improving performance or reducing costs are essential for survival in today's business world. However, innovation can become expensive if uncontrolled. Change for change's sake cannot only cost more than its worth, but it can interfere with normal operations, reducing productivity. Therefore, a systematic change proposal, evaluation, and implementation procedure is needed to minimize wasting time and money on poorly analyzed changes that can't provide the results promised by the originator. Regardless of the system used to evaluate changes, the following criteria should be employed:

1. If the near-term profit from the change is marginal and will continue to be marginal for the foreseeable future, don't accept it.
2. If a high potential for loss exists but a high potential for gain also exists, use a pilot program so that the losses will be minimized if a failure occurs.
3. Don't accept intuitive improvement ideas without objective support. It will force the originator to study his ideas more thoroughly and will reduce management's wasted time reviewing unsuitable recommendations.
4. Innovations should be acceptable to the profession, staff, management, other affected departments, and customer.
5. In general, the implementation of a change will cost and take longer than estimated.
6. Be sure that all aspects of the change have been considered; e.g., labor, space, time to implement, staff cooperation, availability of special equipment, and special skills needed. Also consider need for additional support group labor and documentation requirements.

1.5 Applicability of Book To All Size Companies

At a time when products and equipment are becoming increasingly complex, the small and medium size company trying to meet today's challenges cannot depend on trial-and-error methods. The rationalization that things can be done in a less formal or random way because the company is small is almost certain to lead to failure in today's environment. To survive, small and medium size companies have to adopt the well-defined organization and systematic operations of large, successful firms.

Why are systematic and well-disciplined operations so important for small and medium size companies? Because of the following reasons:

- Change is rampant in most areas of our industrial world; stable, proven products do not necessarily endure and all size companies are forced to innovate; e.g., improve basic designs, add safety features, create completely new products, or use new materials and tools.
- Customer requirements vary; in some cases the customer not only wants to buy a product but to control and monitor its initial design, development, testing, and production. This requires a formal system of operation to meet his needs.
- Tiny, seemingly inconsequential parts can create a disaster if they are unsafe or function poorly when used in an automobile, medical device, airliner, or spacecraft. Therefore, quality and change control are essential to avoid catastrophes.
- The impact of construction and manufacturing activities on the environment, coupled with public and governmental concerns, require systematic techniques for reducing waste, pollution, and destruction of natural beauty and resources.
- The increasing prices of labor, paper, polyester, and other items requires constant vigilance in eliminating waste and improving efficiency.

Regardless of size, the D&D operation is a critical component in the process of designing, developing, building, and modifying products; it plays a major role in minimizing the problems that ensue from the above contemporary conditions and problems. Therefore, it cannot afford to ignore systematic operations or methods for reducing costs and waste.

Of course, new techniques and systems cannot be applied blindly. They must be evaluated not only in terms of a company's size but in terms of its products, cost to implement vs. benefits, customers, and long-term growth plans. For the purpose of this book, small, medium, and large size D&D operations are defined as:

Small: less than 10 people
Medium: less than 50 people, more than 9
Large: more than 49 people

As you read about the techniques and systems described in the following chapters, you will automatically ask: Is this technique practical for my company? Some of the techniques such as computerized status and control systems, are certainly not applicable to a 5-man D&D operation—at least not until computer-aided control systems become much less expensive. However, most of the techniques presented can be applied in modified form to companies with as few as 20 people. To help in evaluating the practicality of adopting these techniques, a checklist is provided in Table II to be used for assessing the worth of the technique to a particular company.

From the checklist, it is apparent that any new system/technique requires systematic scrutiny to avoid installing a new method of operation that does not promise a fair return for your investment. Therefore Figure 3 shows a new technique evaluation summary to help in making a more objective cost vs. benefit analysis. Although many of the items must be estimated by an experienced engineer, designer, or draftsman, the systematic analysis of the cost savings will yield much better results than an intuitive estimate of whether the system will save or cost you money. Besides, even if one is good at intuitive estimates, the evaluation summary should be used to confirm them.

The following procedure is recommended for evaluating a new technique/system:

1. Review the checklist for "yes" answers. At least one of the questions should be "yes" to justify further evaluation of the technique.

Table II

CHECKLIST FOR APPLICABILITY OF NEW TECHNIQUE TO YOUR COMPANY

- Will the cost of implementing the new system/technique be offset by the cost savings derived?
- Can I simplify the technique or reduce its scope of application to minimize its cost but still obtain the benefits needed?
- Can I use the system, with minor modifications, to do more than meet one specific need and thereby increase the benefits derived to justify its implementation?
- Is the risk to product safety and customer dissatisfaction dangerous to the company's future if I don't implement the system, regardless of whether I can justify its cost in terms of cost savings?
- If the technique is not practical at this time for my company, does the anticipated

long-term growth indicate that the technique will be needed in a year, two, or five? If it will be needed in a year, it is probably wise to start implementing it now.

- Based on our present mode of operation, the technique is not needed. However, will I need it in the future? If the business mix changes, new customers are acquired, or new products developed, the technique may become a necessity. In this case, a synopsis of the technique should be placed in an active file for regular review to assure that it will be available when the environment changes.
- Will the new system free an engineer, designer, or draftsman from routine work so that he can apply his technical expertise to more demanding or creative tasks? Will the improvement in morale indirectly increase the output of my company to offset the cost of implementation?

Prepared By: ______________
Date: ______________

Description of Technique

Reason for Consideration

Cost To Implement

Task	Hrs.	Task	Hrs.	Other Costs	Dollars
Research	____	Training	____	New Equipment	________
Design	____	New/Altered Procedures	____	Facilities	________
Coordination	____	Consultant	____	Supplies	________
Review/Evalute	____	Clerical	____	Materials	________
Implement	____			Tools	________
				Travel	________
				New/Altered Forms	________

Total Burdened Labor $________
Total ODC $________
Grand Total $________

Probability of Success (Profitable): ______________________%

Potential Benefits

Reduced Turn Around Time: $____*

Reduced Labor (__Hrs.x$15/hr.) $____

Reduced Materials (Qty. per yr) x ________(Unit Cost): $____

Reduced Equipment ($__cost + $__Maintenance): $____

Reduced Facilities (__x$.75/ sq. ft.): $ ______

Reduced Consultant ($__/hr.x ______ hr.): $______

Reduced Travel/Other: $______

Net Savings Over 1st Year = $____; Next 5 Years = $_______

Approved By: ______________________________ Date ________

Approved By: ______________________________ Date ________

Rejected By: ______________________ Date ________

Reason for Rejection:

*(Estimated Savings due to Fewer Delays)

Figure 3 New Technique/System Evaluation Summary

2. The new technique/system evaluation summary is prepared next to verify the value of implementing it.
3. Review of the evaluation summary with your key people to be sure all areas have been examined for possible problem areas, unrealistic estimates, and overlooked items.
4. If the technique passes these tests and is accepted, assign one person to evaluate the best approach, design forms, write procedure, and implement system.

It is emphasized that a random approach to most operations is more expensive than a systematic one defined in writing and approved by management. Therefore, when a new technique is implemented, it should also be documented and followed up to be sure that it is working as designed and providing the desired benefits.

2

2.0 Design and Drafting Department: Key to Systematic Operations

The design and drafting (D&D) department is responsible for preparing most of the drawings and associated engineering data needed to depict ideas and designs created by the engineering department. Efficient preparation of engineering data requires a balanced mix of draftsmen, designers, checkers, equipment, supplies, facilities, and techniques. This chapter is devoted to reviewing these areas in terms of the D&D department's overall operation and performance.

In preparing engineering documents, the goals of the D&D department should be:

1. Accurate, clear, and complete preparation of data at minimum cost and labor.
2. Rapid turn-around of data to originator to minimize delays in its progress and that of its users.
3. Technical assistance and recommendations for improving the designs submitted by engineers and scientists such as techniques for design simplification, cost reduction, and improved reliability or safety.
4. Detection of errors in designs submitted.
5. Use or development of new techniques, materials, and instruments for increasing the department's productivity and for reducing operating costs.

6. Complete and accurate identification for each document by drawing number, title, project, and customer.
7. Use of standard procedures for content, format, style, and revision of data.
8. Completion of work within cost estimate.

BENEFITS AND APPLICATION OF THIS CHAPTER

This chapter provides an overview to the D&D operation. Its prime benefit is in highlighting areas that are of special concern in this operation. Benefits include:

* Gives guides for an appropriate organizational structure, depending on size and needs of company.
* Provides a review of three types of organizational approaches and the advantages and disadvantages of each.
* Provides practical information on various aspects of the D&D operation such as salary ranges, physical layout considerations, and educational/experience levels for different categories of draftsmen and designers.
* Gives a review of various techniques for controlling D&D activities and performance.
* Gives sources for additional information on graphic technology.
* Presents techniques for reducing operating costs, improving efficiency, or obtaining better performance.

The material presented in this chapter is applicable to all sizes and types of organizations, depending on the needs and problems of each company. If the applicability of a technique is questionable, refer to Section 1.5 for evaluating its benefits.

2.1 Organization and Staff: Principal Elements for Superior Performance

Three basic forms of organizations are available for D&D personnel. These are line, project, and matrix. Figures 4 and 5 show these types of organizations.

A line organization is one that has its D&D members report directly to a D&D manager, who is responsible for hiring, firing, and total staff performance. This type of organization is also called a centralized department.

A matrix organization is a combination line and project organization. Draftsmen and designers are assigned to one or more projects and follow the directions of the project managers when working on the project. However, they are still directly responsible to the D&D manager, who determines their pay raises, controls their assignments, directs their training, and sets standards.

In smaller companies, a formal department called D&D may not exist. Instead, the D&D staff reports to a chief engineer or principal architect. Under this type of line organization, engineers, designers, and draftsmen report directly to one head man. However, with such a small group, the D&D staff works closely with engineers to get the drawings done. When there are several draftsmen in a small company, a lead draftsman may be used to schedule and assign work.

A project or team organization consists of D&D personnel who are assigned to specific projects as they arise. The D&D personnel report directly to the project manager and are essentially disconnected from the parent D&D department until the project is over. When the project is completed, these individuals return to the department for reassignment to another project or to do

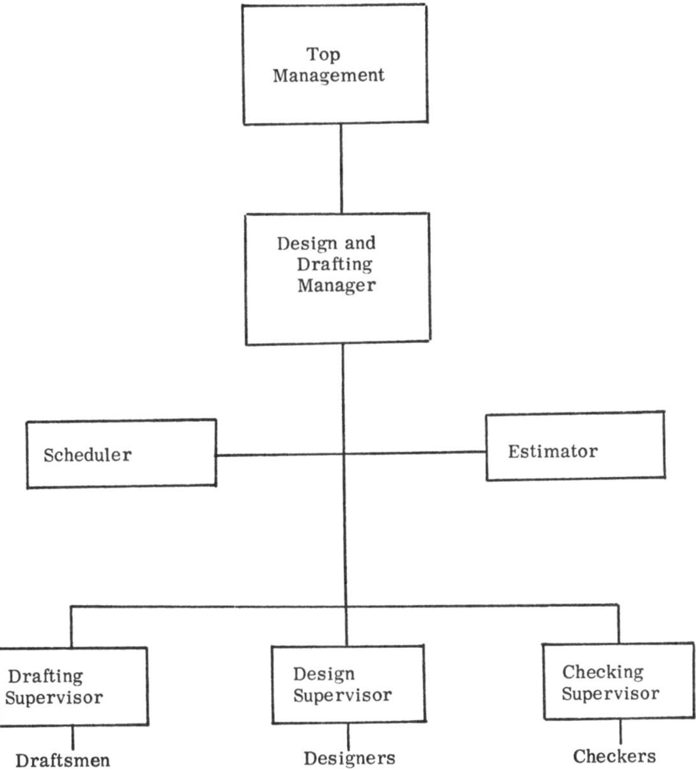

Figure 4 Line Organization

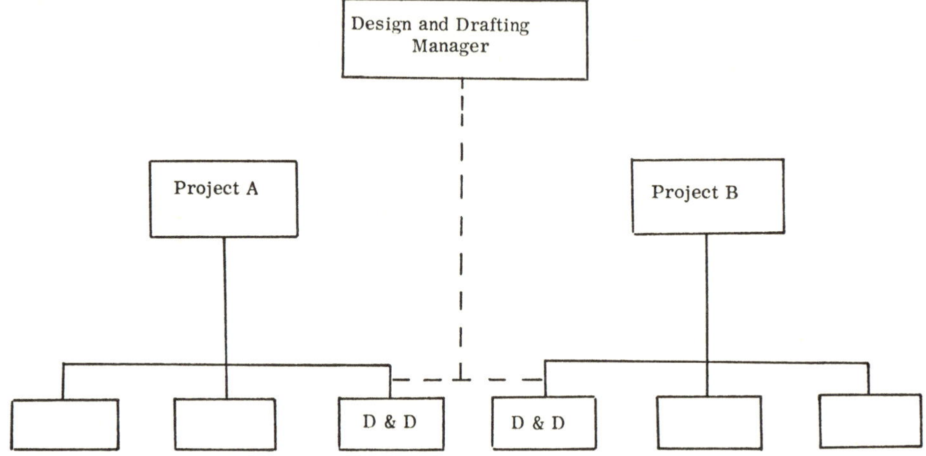

A. PROJECT ORGANIZATION

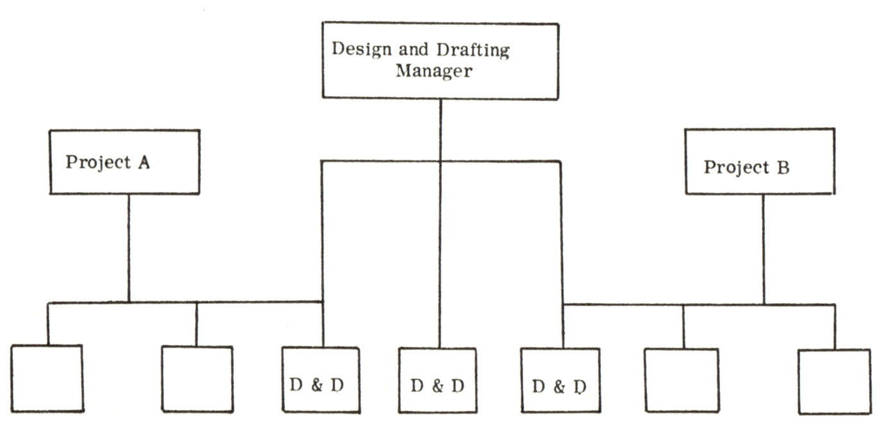

B. MATRIX ORGANIZATION

Figure 5 Project and Matrix Organizations A. Project Organization B. Matrix Organization

departmental work. Therefore, the D&D department is essentially a holding operation for coordinating activities and assigning personnel work until a new project assignment is made.

The advantages and disadvantages of line, project, and matrix organizations are given in Table III.

Because the line type of D&D organization lends itself to a more systematic description of D&D operations, this type of organization is discussed next.

Table III

Advantages and Disadvantages of Different D&D Organizations

Type	Advantages	Disadvantages
Line/ Centralized	More effective control and use of staff.	Project authority and control is weak.
	Concentration of specialists increases use of collective experience and knowledge.	Strongest groups or project leaders outside D&D tend to get better service.
	Standardized hiring, operations, and drafting criteria improve overall efficiency.	Responsibility for project progress and cost is difficult to pin point.
	Better development and application of new drafting/design techniques.	Decisions on problems or needs related to a specific task or project take longer to make because the manager is concerned with many projects.
	Top flight manager can be recruited and rewarded for improved performance of the department.	Unless well-managed, work output per person may be less than on a project basis.
	More opportunity for professional growth.	
	Greater versatility and flexibility.	
Project/team	Well-suited for tight organization and control and short schedules. Provides rapid reaction time.	Continuity of team personnel and technical developments will be poor unless teams can be immediately reassigned to a new job.
	Staff for project is hand picked for project based on experience and past smooth interaction with project team.	Technical exchanges are poor among D&D personnel on different project teams.
		Less stable work is available for personnel.
	Strong control by one person having one objective to meet.	Personnel tend to get more involved in each others problems. In general, this dissipates energy which could be used more profitably in their own areas.
	Successful teams develop strong espirit de corps.	More difficult to time the transfer of staff to other work as one project ends and another begins.
	Better cost control because spreading of time among several projects is not required.	Growth/promotion possibilities are more limited when teams are dissolved after each project.
		Harder to balance work loads; e.g., personnel on Project 1 may be working 6 days a week while members of Project 2 are busy only ¾ the time.
Matrix	Provides better project control than obtained from a line organization.	More conflicts over priorities and assignments between department and project managers.
	Provides more stable and continuous employment for staff than project organization.	Staff members are accountable to two bosses who may have conflicting demands and goals.
	Offers better exchange of information among projects than with project organization.	Accountability of profit and loss is more difficult; e.g., project and line managers blame each other for cost overruns or performance problems.
	Provides better continuity and use of technological developments when compared to pure project organization.	

A line organization chart for a D&D department is shown in Figure 4. The design and drafting manager is responsible for the entire operation and has a staff consisting primarily of designers, draftsmen, and checkers. When the department is large, a scheduler and cost estimator may also be added to the department in staff positions. However, these functions are usually assigned to the design and drafting supervisor.

While several variations in the department's organizational structure exist, the principal feature should be assignment of independent responsibility for the entire department's performance and cost control to a D&D manager. As shown in Figure 4, the D&D manager reports to a higher level manager such as an operations manager, engineering manager, or technical services manager.

As mentioned, the staff of the D&D department includes the manager, scheduler, cost estimator, draftsmen, designers, and checkers. The extent of the staff and number of supervisors depends on the size of the department. Therefore, all the staff members described next may not exist in a medium or smaller-size department. However, someone must perform their functions.

TOP MANAGER: The function of the top manager is to monitor the overall performance of the D&D department and to work with the D&D manager to obtain the funds, facilities, equipment, and personnel needed to provide the documentation services required by D&D and other departments. His high position in the company provides him with the leverage to obtain needed funds or policy changes, which the D&D manager may have more difficulty in getting on his own.

Another function of the top manager is to communicate information to the D&D manager about the latest changes in company objectives, growth plans for future, new documentation requirements, new product lines, and feedback from other departments on the D&D department's performance. (Dissatisfaction with certain services may not always be given directly to the D&D manager; therefore, the top manager should give this information to the D&D manager as soon as possible.)

D&D MANAGER: The D&D manager is responsible for the overall performance of his department, including cost and schedule compliance. In addition, he is the prime source of innovation and makes changes to improve performance, quality, productivity, and capabilities.

Because of his responsibility for the department's overall performance, he is a key contributor to the preparation of the design and drafting manual and should approve all changes to it. This manual, which is discussed in Section 2.5, defines the rules, procedures, and standards to be followed by the staff in performing its work.

Another function of the manager is to solve problems. Whenever technical, administrative, or personnel problems arise, he must become thoroughly involved until they are resolved.

Continual monitoring of the operation is essential if gradual organizational decay is to be avoided. A concerted and consistent effort is needed to assure continued high performance from an effective department. It cannot be assumed that once the department reaches a certain high level of output that it will automatically maintain this level by itself. Work must be constantly performed to realign displaced or poorly performing components on a real-time basis to avoid gradual deterioration. For example, periodic reviews of the actual method of revising released documents are needed to be sure they conform with existing standards. When discrepancies exist, their cause can be determined and corrective action taken.

Training is another responsibility of the manager. He must set up a program for instructing new or experienced people in new techniques. Periodic review of the D&D manual is one way to keep people up-to-date and to discover discrepancies between the manual and real-life operation. If real-life operations are indeed better, the manual should be revised to reflect this. Of course, anytime a new operating procedure is released, the staff must be thoroughly indoctrinated in its use. A new system, no matter how sound, is useless unless the staff understands it and is convinced of its value to the department and company. Note that although companies spend considerable effort devising new operating systems, they often fail to follow-up with adequate effort on their implementation.

Obtaining facilities, equipment, materials, and personnel for the department is a continuing activity of the D&D manager. He must have the persuasiveness of an effective politician to convince management with well-prepared, sound proposals that his needs must be met for the welfare of the company as well as the department itself.

DESIGN SUPERVISOR: The design supervisor is responsible for directing the work of his design team and maintaining high quality and productivity. He coordinates work assignments, keeps track of expenditures in his area, and develops improved methods of operation for his team. When a cost estimator and scheduler are not available, he assists the D&D manager in these areas. The design supervisor also works closely with the manager to assure that department objectives, policies, and procedures are being implemented.

Other functions include setting design standards and reviewing layouts, designs, and calculations for accuracy and compatibility with project requirements.

DESIGNER: The designer is an experienced draftsman or engineer who has specialized in design and drafting practices. His creative ability and experience allow him to convert sketches, conceptual drawings, and engineering calculations into design layouts of the end item. He also knows what specific components can be purchased for use in the end item and can assess the manufacturing feasibility of the design. His wide experience also enables him to help determine the cost to manufacture a part or end item. In addition, he has a good understanding of engineering and design principles related to his area of specialization.

In electronics, he has special skills which enable him to do detail layouts of components on printed wiring boards. This area requires expertise in specialized electronic parts and components, printed wiring techniques, and electrical and mechanical requirements.

A mechanical designer specializes in the ability to visualize mechanical forms and to develop feasible configurations and structures that can withstand the physical stresses that the structure will be subjected to. He is also familiar with various types of materials, finishes, and manufacturing techniques. Other functions include checking computations and assuring compliance to company standards.

Designers are often categorized as A or B to designate level of experience. The A category represents the more experienced level.

DRAFTING SUPERVISOR: The drafting supervisor performs similar functions to those of the design supervisor described before, except that he directs the work of draftsmen. When a design supervisor is not needed, the drafting supervisor oversees the work of all designers and draftsmen.

DRAFTSMAN: The draftsman works on routine document preparation. His main responsibility is to translate engineering sketches, design layouts, and notes into finished detail documents that conform to standard drafting practices. He is not responsible for the technical accuracy or content of the original document submitted to him. However, as he gains in experience, he can do the following:

- Detect omissions and errors and obtain additional information from the engineer or designer.
- Question the validity of the technical details and recommend improvements.

The skill level of the draftsman depends on his natural talents, education, and experience. He should be familiar with the drafting manual, drafting techniques, dimensioning methods, revision procedures, and identification requirements. In addition, he should know where to find information when needed.

Draftsmen are usually placed in three grades such as Draftsman A, B, and

C, with C being the lowest skill level. Design draftsman is a fourth category that lies between a draftsman A and a designer.

CHECKER: The checker verifies that the draftsman's or designer's work matches the original rough document in content and accuracy. He also checks for feasibility of manufacturing and inspection and for conformance to standard drafting practices. A checker is a very important part of the D&D group because he can save the company many thousands of dollars by catching errors before hardware is built and thus avoiding the scrapping of defective parts. The checker may be a senior draftsman, a designer, or an engineer.

A senior level checker can offer services beyond catching normal drafting errors. He can detect errors in design and recommend improvements not recognized by the engineer.

SCHEDULER: The scheduler processes all work from other groups to the D&D staff. His function is to feed work into the department at the best rate possible and in the order of importance. He should not hold up a rush job for a routine one just because it came in after the routine job. If there is any doubt, he can check with higher management to have a priority set in terms of overall company benefits.

When work loads seem excessive and delays of important work seem likely, he must make arrangements to subcontract the work out or obtain job shoppers until the overload is eliminated.

Another function of the scheduler is to follow-up on work in progress and to notify the originator of any problems. On extremely tight schedule work, he may have to use short interval scheduling (see Section 2.7) to obtain the desired response.

Note that many companies have the D&D manager or drafting/design supervisor schedule work. Therefore, a full-time scheduler would normally be used for a large department or where it is found more efficient than other methods.

ESTIMATOR: The estimator is responsible for determining the cost to prepare an individual drawing or a complete document package for a project. Accurate drafting estimates are difficult to make. Therefore, extensive experience and persistent collection of data on documentation requirements are needed to estimate drawing costs.

CASE IN POINT: Failure of a medium size company to use a professional estimator in bidding a contract requiring high quality microfilm and aperture cards of drawings resulted in a loss of $3,000. This situation occurred because an experienced estimator was not available and the chief draftsman didn't understand the scope of the effort. He therefore esti-

mated the microfilm cost at $500. Two years later, the company wound up paying over $3500. This is a fairly common experience in industry because adequate manpower of the right kind is not available to investigate the real cost of the work needed to meet contract requirements.

Another key function of the estimator is to keep cost records for comparison with his estimates so that they can be corrected when he is being consistently low or high. He also keeps data on changes in scope of work and error rates to account for overruns.

For many companies a separate estimator may not be practical and this function can be handled by the design/drafting supervisor. In other companies, the D&D manager may estimate data preparation costs unless he has a designer or senior draftsman who is competent in this area.

SECRETARY/CLERK: A secretary or clerk is necessary to provide an interface between the department and other departments for routine communications. In addition, she/he can type reports, keep files up-to-date, and perform miscellaneous duties, which would otherwise be done by higher paid draftsmen or designers.

D&D EDUCATION AND EXPERIENCE REQUIREMENTS

Education and experience guidelines for various D&D personnel are summarized below. For a detailed chart refer to "Standardizing Drafting Job Descriptions," 1971, published by the American Institute of Design and Drafting (AIDD).

Classification	Education	Experience (yrs)
Manager	2 yrs college/or more	10
Supervisor	2 yrs college/or more	8
Designer	2 yrs technical school	6
Design Draftsman	2 yrs technical school	5
Draftsman A (Senior)	—	3
Draftsman B (Draftsman)	—	2
Draftsman C (Junior)	1 yr technical school or	1

Note that four grades of draftsmen are covered. For detailed definitions of D&D job descriptions refer to the document given above. See Section 2.7 for address of AIDD.

Salary ranges for D&D personnel are given below. These are nominal salaries and depend on location, company, industry, and year.

Classification	Hourly Range Within Classification
Manager	$7 to 12
Administrative assistant	$6 to 10

Supervisor	\$6 to 10
Checker	\$6 to 9
Sr. Designer	\$6 to 9
Designer	\$5 to 7
Design draftsman	\$4 to 6
Sr. Draftsman	\$3.50 to 5
Draftsman	\$3 to 4
Sr. Clerk	\$3 to 4
Jr. Clerk	\$2.25 to 2.75

ORGANIZATIONAL VARIATIONS CAN IMPROVE OPERATIONS

The aim of the centralized organizational model depicted in this chapter is to illustrate the key functions performed by the D&D department. It is not meant to prescribe an official guide. Most companies have D&D organizations that are designed to fit their needs and the experience and preferences of their D&D managers. For example, one company may use a special group for developing new operational techniques that increase the department's efficiency. Other companies may have the reproduction or data control departments under the D&D manager's direction. Therefore, selection of the best organizational structure depends on the individual company, operational constraints, and other considerations. Some of these considerations are listed below:

1. Types and scope of drafting efforts; for example, frequent need for short-run, tightly controlled projects or a steady work load of fairly standard drawings.
2. Types and levels of people available. Some people provide greater effort and quality on a project basis than on a line one; others, however, find that a centralized organization brings out the best in them and promotes their efficiency.
3. Size of organization.
4. Stability and continuity desired.
5. Desire and need to develop new operational techniques.
6. Management attitudes toward the D&D operation.
7. Past success of a particular organizational type.

2.2 Facilities, Equipment, and Environment For Getting The Job Done Efficiently

The facilities and equipment needed for the D&D operation are, of course, dependent on the type and size of the department. However, regardless of its size, certain basic items are needed for any department such as drafting and reference tables, desks, vellum file cabinets, and supply cabinets. A sample list of items follows:

Furniture and Equipment

- Drafting tables and stools (one set per designer/draftsman)
- Drafting machines (one per table)
- Reference tables
- Light tables
- Table-model whiteprinters
- Contact printer/vacuum frame
- Office copier
- Storage cabinets
- Flat file for inprocess documents
- Book shelves or cases
- Regular files for records
- Rotary desk files
- Tubular files
- Paper cutters
- Typewriters (regular and for drafting use)
- Automated drafting machines
- Microform viewers

Instruments and Tools

- Pantographs
- Leroy lettering sets
- Planimeters
- Lettering and ruling pens
- Symbol and lettering templets
- Pencil sharpeners
- Technical fountain pens
- Slide rules
- Erasing shields
- Pencil file and tack lifter
- Electric eraser
- Straight edges
- Protractors
- Proportion scales
- Parallel rules
- Adjustable triangles
- Steel eraser
- Refillable pencil pointer

- X-acto knife
- Drawing pencils, leads, and crayons
- French, ship, and radius curves
- Triangles, T-squares, and line spacers
- Flat and triangular scales
- Scribing and graving instruments for map making
- Proportional dividers
- Bow pens and pencils
- Dividers
- Compasses: bow, beam, and drop bow
- Paper shears
- Dust brushes

Materials

- Vellums preprinted with border, title box, company name, etc. (Sizes: A, B, C, D, etc.)
- Tracing paper with grid lines
- Graph charts
- Whiteprinter sensitized paper and film—various sizes and types
- Polyester film
- Artboard
- Polyester gridline sheet for tracing guide
- Ink, stamp pads, and note paper
- Rubber cement
- Drafting powder
- Mending tape
- Masking tape
- Transparentizer
- Ink and brownline eradicator
- Lead cleaner
- Rub on letters, symbols, and ribbon
- Opaquing liquids

DRAFTING ROOM: A GOOD ENVIRONMENT MEANS BETTER PERFORMANCE

The D&D environment can affect efficiency and morale. Therefore, the layout, conditions, and decor of the facility should be carefully assessed to optimize performance. The space per individual, for example, can affect comfort, free movement, and productivity. An average of 100 square feet per person is a desirable working space. (Many companies, however, operate with as little as 50 square feet per person.)

Besides improving morale, pleasant working conditions reduce fatigue, irritability, and promote greater productivity. For example, it has been found that workers get irritable when sheltered from natural daylight. Some studies indicate that productivity can increase by about 8 percent with a good working environment.

Environmental factors that should be evaluated include lighting, noise, colors, temperature, humidity, and space. Recommended values are listed below:

* Light: 200 ft.-candles[1] for details, 150 for layout work
* Noise: much less than 75 db[2] Notes: noise level in average office is 55 db. Carpeting can reduce noise by as much as 50%.
* Temperature: 63 to 72 degrees F
* Humidity: 30 to 70 percent
* Air flow: about 1000 ft^3/hour/person
* Colors: cheerful and bright

Layout of the facility is also important. A checklist for arranging the department is given in Table IV.

Table IV

Checklist for a D&D Department Layout Requirements

- Main aisles should be about 5 feet wide; secondary aisles should be 3.5 feet. Avoid having drafting tables facing main aisles. When necessary, however, be sure to allow extra space (about 2 feet) for extension of the drafting machines during use.
- Increase width of aisles when filing or storage cabinets open into aisles. This space is needed to accommodate open file drawers and cabinet doors.
- If drafting tables are lined up one in front of another, the working space between them should be greater than 3 feet. If more than two tables are in a row, add another ½ foot to allow the inside draftsman to get in and out without bothering the other draftsman.
- Arrange drafting tables in the same direction so that each draftsman faces the back of the man in front of him. Face to face positions lead to wasted time due to distractions.
- Seating space between desks should be about 2.5 feet wide.
- Place people who have frequent visitors near the entrance door. This also applies to staff members who are in and out a lot.
- Personnel should be within a few feet of the files and equipment that they normally need.
- Eliminate unnecessary furniture from active work areas.
- All D&D personnel working on related projects should be grouped together. Also put people as close as possible to the person responsible for them.
- Drinking fountains, vending machines, and restrooms should be convenient to use. (A 1-minute roundtrip to a restroom will cost the company about $3000 per year if 20 men make 3 trips a day.)
- Supply enough telephones so that time is not wasted walking to and from distant locations or waiting to get on the line.

- Whenever feasible, place files against walls because they take up less space that way. Use lateral files when aisle space is limited.
- Provide reference tables for those who need them and as close as possible to their drafting tables. (Some drafting sets come with matching reference tables.)

2.3 Design and Drafting as a Service Group

The D&D department is responsible for serving engineering and production. To do this job well and quickly, the D&D personnel must not only provide needed graphic services but must minimize red tape and rough interfaces. For example, a request for drafting services by engineering should be as simple and straight forward as possible. The design engineer doesn't want the delay and trouble of filling out a complicated form and then obtaining several signatures before he can submit his sketch or layout to D&D. Therefore, any forms or signatures required should be essential for effective control. In addition, a D&D clerk or draftsman should complete and process the paperwork required with a minimum effort on the part of the engineer. This procedure may increase D&D labor costs but these costs will be more than offset by engineering labor reductions.

Other key points in providing a good service are responsiveness and cooperation. Engineering or manufacturing are often unrealistic in their requirements or behind schedule because of unforseen problems. As a result, D&D's schedule is shortened to make up for the lost time and to prevent the project from slipping. Although this places an unfair burden on D&D, complaints and feelings that other departments don't appreciate the time required to do good graphic work should not be allowed to result in a poor response to the challenge.

Of course, a compressed schedule may require more funds because of overtime or additional manpower at higher labor costs. However, whatever is needed to meet the schedule should be done if the quality and accuracy of the data are not compromised. Of course, the expenditure of more than allotted funds must be approved by the project or engineering manager.

If engineering is consistently cutting short the allowable D&D time for a task, then a meeting should be held between department heads to discuss the problems and to determine how to correct them.

2.4 Control Techniques For Lower Cost and on Schedule Completion of Work

Certain minimum control techniques are needed to attain an efficient D&D department. However, keep in mind that the scope, sophistication, and frequency of control techniques depend on the size and complexity of the department itself. For a small department, too much control can be as expensive as too little.

DRAFTING WORK REQUEST PROVIDES IMPORTANT INFORMATION

A drafting work request (DWR) is the official basis for authorizing D&D to perform work on a project. It should be prepared by each engineer submitting work to the D&D department. The DWR form identifies the work to be done, who it is for, the charge number, when it is needed, and other pertinent information. Although all entries required need not be completed by the engineer, certain information that only the engineer knows must be recorded.

The scheduler or manager accepts this request and the rough form documents. He then makes an estimate and either notifies the requester for approval or passes the form with the estimated drafting time to the drafting supervisor or draftsman. The draftsman works on the drawing; he records his hours on the form. After completion, he returns the form, completed drawing, and original rough sketch to the scheduler. If checking is needed this is done before the new drawing is returned to the requester for final review, approval, and release for use.

The completed drafting work request is presented to the secretary or filed by the scheduler for future reference.

Besides acting as a legal authorization to expend funds, the DWR helps control the quality, scope of work, and identification of the drawing by providing the draftsman or designer with a compact summary of the work to be done. It also allows his supervisor to verify that the draftsman is doing what was wanted and how much labor was expended. A sample DWR is shown in Figure 6. A unique number should be assigned to it for control and accounting of drawing charges.

MONITORING THE PRODUCT WITH PERIODIC DESIGN REVIEWS

Design reviews have several important functions such as to provide management with data for determining the product's status, identifying problem areas, approving production of hardware, conducting trade-off studies, and detection of undesirable parts.

CASE IN POINT: A new product was the subject of a design review. In reviewing the materials and parts required to build

the product, one engineer that had used the same part in a previous program identified it as an item that the vendor could not deliver in the short time required by production schedule. On his previous program, he had received the parts 2 months after the promised delivery date.

As a result of this information, another part was selected from a vendor with a reliable record of meeting promised delivery dates.

The net savings to the company in reduced costs due to avoiding production delays and lower sales for the period was several thousand dollars. (Similar savings can be obtained from design reviews by detecting and replacing defective parts discovered in other products or programs.)

DRAFTING WORK REQUEST No. ________

Project Name ______________________ Contract/Job No. ________

Requester __________ Title __________ Dept. __________

Suggested Title of Document ______________________

Next Higher Assembly Name and Number

Description of Work

Date Submitted to D&D ______________ Priority Level ________

Date Work Must Be Completed ______________ Quality Level ________

Documents: Applicable (Required) Reference

Estimated Hours to Complete ______________

D&D Level Required Designer ______________ A Draftsman ________

B Draftsman ______ C Draftsman ________ Other ____________

Actual Hours Required to Complete ______________

Actual D&D Level and Persons Used __________ Name __________

Remarks/Problems

Approved:

______________________ ______________________

D&D Manager Date

Figure 6 Sample Drafting Work Request

Requests for changes to unrealistic specifications, and establishment of test programs are also products of design reviews. Other functions include identification of key documents defining and controlling the product, review of basic design approach for weaknesses and impractical manufacturing requirements, and design change control. In summary the principal objectives of design reviews are:

1. To achieve the best design approach possible by critically reviewing all major electrical and mechanical aspects of a product.
2. To identify and confirm the final design and quality base for building end item.
3. To avoid errors in design and fabrication found in other similar equipment already built.

Design reviews should be held at key points in the progress of a project to define baselines from which engineering and production changes can be referenced. By doing this, each stage of the design or baseline can be used to set more stringent design change criteria to minimize the impact of changes. For instance, as the design progresses, each baseline represents a departure that requires more stringent control of changes to avoid excessive delays and expenses resulting. Typical baselines include:

- Conceptual—basic product idea or configuration is developed
- Preliminary—preliminary engineering design is completed and evaluated
- Critical—detail drawings are completed for building a prototype
- Production—production drawings are released and manufacture of product begins

Each of these baselines is identified by drawings, specifications, parts and wire lists, test procedures, and other technical data. However, the level of detail and explicit nature of these documents increases as the product design progresses from the conceptual stage (minimum and tentative data) to the production stage (complete and final data).

Design reviews can be simple affairs requiring only a few hours or they can be complicated operations covering many days. The engineering department is usually responsible for organizing and implementing the review. Although D&D personnel participate in the review, they are only one group of many. For example, participants may include design engineers, product managers, manufacturing engineers, quality assurance and reliability engineers, procurement representatives, and field engineers. Review drawings should be sent to all interested people for comments or approval. Approval may be assumed if comments are not received by a given date in some cases.

NUMBER ASSIGNMENTS MUST BE CONTROLLED

The drafting supervisor is responsible for assigning a number to each drawing prepared by the department. Because it is extremely important to have each document marked with a unique number that is not assigned to any other document, one person should be responsible for this task. A log book should be maintained so that the drafting supervisor can assign a number and record its title, project, and the date in the book before releasing the drafting work request to the draftsman for preparation of the document.

SIGN-OFF CYCLE ENSURES PROPER EVALUATION OF DATA BEFORE RELEASE

Before releasing an engineering document to production, test, or inspection, it should be reviewed and approved by the right people. Normally this involves the originator, the draftsman/designer, the checker, product engineer, manufacturing engineer, and quality assurance representative. The drafting supervisor should verify that all responsible people have reviewed and signed the drawing release blocks in the title block of the document. The objective of the sign-off process is to assure the accuracy and completeness of the document. Therefore, only people who can contribute to satisfying this objective should be in the cycle. Otherwise, additional delays and wasted labor will result.

If a drawing is to be used for non-production or experimental uses, only one or two key people need sign the document before it is released to data control for reproduction and distribution. For example, the signatures of the originator and draftsman are usually sufficient for release.

Upon completion of the sign-off cycle, a release and issue form should be prepared by the drafting supervisor and submitted to data control with the original document. This document authorizes the clerk to issue prints to the staff. It also provides important supplementary data such as next assembly numbers, type of release, and effectivity.

STORAGE OF INPROCESS VELLUMS PREVENTS LOSS OR DAMAGE TO VALUABLE DATA

Inprocess vellums need protection just as released ones do. Therefore, the D&D department should have a lockable file for storing in-process documents at the end of each day or week. In some companies, locking up vellums may not be necessary but if classified or proprietary drawings are being prepared, they must be secured before the draftsmen leave for the day.

CHANGE CONTROL PREVENTS UNNEEDED DESIGN ALTERATIONS

If the configuration of items depicted on drawings and other engineering data is to be assured as that officially approved, then released vellums must be protected against unauthorized alteration. Therefore, a formal system of change processing is necessary. This usually involves a change authorization document that describes the change and also contains signatures of key people required for making a change to a vellum or simply attaching an approved engineering order to the released print (see Chapter 6).

A design change notice is prepared by a designer or draftsman and approved by the staff to obtain a vellum to be changed from data control. After revision, of course, the new document is re-approved and re-released to data control for printing and distribution.

PROGRESS REPORTS KEEP MANAGEMENT INFORMED

For a small department, monitoring of progress can be obtained simply from on site inspection and verbal reports. For larger departments, formal weekly reports may be required from each supervisor. Regardless of the method of reporting, a regular flow of status information to the D&D manager is necessary for him to keep abreast of problems and progress so that he can intelligently report to his boss. Note that status information applies to cost as well as to output and schedule data.

Depending on the department's operation and staff, reports may be limited to exceptional events rather than to routine ones. Thus the manager is flagged when action is required on his part. Of course, the manager may not be satisfied with this type of reporting if he wants to keep on top of all the detailed activities of his department.

SPECIFICATIONS, STANDARDS, AND MANUALS: GUIDES FOR CONSISTENT DATA

Specifications, standards, and manuals provide requirements to follow in preparing documents. They can define acceptable drafting practices, abbreviations, methods for identifying products, and microfilming requirements. Thus, by calling out appropriate documents for a project in the drafting work request, specific controls can be established for preparing engineering documents.

SCHEDULES IMPROVE PERFORMANCE AND PLANNING ACTIVITIES

Schedules are key control elements. They provide a visual method for keeping track of progress on a daily, weekly, or monthly basis and for determining actions required to expedite work. In addition, their preparation helps assure that a project has been broken down into its basic tasks and that their completion times are compatible with the overall work period. A discussion of how schedules are prepared is covered in Chapter 10.

2.5 Design and Drafting Manual: A Key Standard For Consistent and Economical Work

Drafting manuals are a part of the control mentioned in Section 2.4, but are presented separately because of their special importance. A D&D manual presents policies, procedures, standards, and basic information related to the department's operations. By establishing management approved systems and related data, it helps assure uniform and efficient production of engineering documents.

The instructions and policies covered in the manual should be strictly enforced. However, temporary deviations may be permitted to meet customer or contract requirements or to adapt to special situations. Permission from the D&D manager or person responsible for the manual should be obtained in writing. The period allowed for this deviation, the specific practices that will be modified or not followed, and the reasons for the deviation should also be documented.

Drafting manuals can reduce drafting time, errors, costs, and waste when carefully prepared and followed. In particular these documents offer these advantages to the D&D group:

1. Provide minimum quality levels for different types of documents.
2. Provide accepted and uniform practices in the preparation, use, and maintenance of engineering data.
3. Reduce costs that occur when drawings have to be reworked because they don't meet minimum acceptable standards.
4. Eliminate conflicts between parties with different backgrounds and ways of doing things by providing one way to do it that is sanctioned by management.
5. Provide draftsmen instructions on how to do routine tasks without requiring the assistance of the drafting supervisor or manager.
6. Provide supplementary and useful D&D technical information.
7. Serve as training aids.

Another practical reason for having a manual is that large customers or the government may require companies to have a manual that is acceptable to them or they will impose their own requirements.

Drafting manuals cover a broad range of topics and therefore are often large documents which may exceed 200 pages. The major topics are identification of data, drawing types, drawing format, drafting practices, dimensions and tolerances, lettering, part identification, change procedures, abbreviations, drafting symbols, tables of standard drill hole tolerances, trigonometry, and bend radii.

The information on these topics may be described in detail or referenced to other documents used as standards. Referencing to other documents is a good practice for detailed and little used information, especially if the documents are readily available. Industry and government have standards and specifications that cover topics which are often used in drafting manuals.

Note that a comprehensive index increases the utility of the manual because it reduces look-up time and accelerates finding the right topic when the draftsman/designer is in a hurry.

The drafting manual should be reviewed periodically (e.g., annually) to assure that it reflects current and improved practices and that conflicting sections have been removed during previous revisions. Failure to eliminate redundant, contradictory, and antiquated practices will defeat its purpose and undercut the staff's reliance on it for guidance.

2.6 Standards Mean Better Quality and Lower Costs

Standards should be established in several key areas. These include format, drafting techniques and quality, cost, operating procedures, identification systems, and job descriptions. Of course, these topics will appear in the D&D manual if the company has one. However, if a manual is not prepared on these areas and applicable reference documents are not available, separate documents should be prepared on as many topics as possible.

Format refers to the layout and standard arrangement of title blocks, revision blocks, and borders used for the documentation being prepared. The exact format of the various types of documents is not important when it meets the needs of the company or user. However, a standard format for each type of document is essential for uniform looking documents and ease of preparation. In addition, a uniform format assures that certain minimum requirements will be met regardless of the draftsman or designer doing the work. Of course, standard formats help reduce waste, confusion, and revision. Because they are standard, the draftsman doesn't have to think much about this area and work progresses faster, reducing costs. For example, if by use of a standard preprinted format, 40 draftsmen each save 5 minutes per drawing and they prepare a total of 2400 drawings a year, the company saves $2400 (minus cost of vellums) per year. (Assuming a burdened hourly rate of $12.)

Standard drawings are also cost and time saving devices. A standard drawing is prepared for a part that has a constant basic configuration but its dimensions vary for different applications. In this case, the standard drawing is used to prepare a vellum for each part and is modified by adding the dimensions for each desired configuration. Using this kind of a drawing can save a small company a few thousand dollars per year and bigger companies a lot more. (Note: standard drawings are also prepared for commonly used items and materials to define their key features.)

Standard costs are also important to an efficiently running organization. Although blind reliance on standard costs for various sizes and types of drawings can lead to serious overrun or underrun problems, some guidelines are necessary for accurate cost estimating and control. Cost estimating and related information are covered in Chapter 9.

Time estimating is another area that requires standardization so that work can be scheduled more effectively. Thus a certain category of drawing on an A size sheet should take X hours to prepare. A B size drawing Y hours, and a D size drawing Z hours. Of course, cost estimates are based on both time estimates and the hourly rates of the draftsmen. Note that time estimates should also include a certain amount of revision or rework time. This rework should be adjusted to the engineer submitting the original sketches. Some engineers will be notorious for constant revisions while others will make only moderate revisions to their designs after release to D&D.

The method and format for identification of engineering documents must be standardized to assure uniform and consistent assignment of titles and document numbers. A disciplined system of identification reduces confusion, prevents identifying more than one document with the same number, and allows cross referencing drawings to expenditure of funds and technical data.

CASE IN POINT: One company created a chaotic situation in its D&D group by not requiring the assignment of numbers to drawings when preparation was first begun. As a result hundreds of in-process drawings were floating around the department without unique identifiers. The result: No easy way of recording time spent on drawings or jobs and difficulty in associating a partially completed drawing with its layout, higher assembly drawing, project, and engineer. To correct this situation the company had to require that a drawing number and title be assigned to each drawing before work was begun on it. This information was recorded in a log book so that the drawing number could be used to cross-reference it to the originator, job, and date of assignment.

Job descriptions are another area for standardization. They provide uniform and well-defined descriptions of various personnel in the D&D department. Thus the duties, education, and experience for each classification are explicitly defined. The American Institute for Design and Drafting had developed seven job descriptions in its publication "Standardizing Drafting Job Descriptions," 1971. Figure 7 shows one of these job descriptions.

In summary, formalizing a method of operation should be done when it will reduce costs, improve reliability of the operation, or improve the quality of its results. Standardized procedures, when properly applied, simplify operations by reducing uncontrolled variations that increase errors and take longer to do.

Classification #3—Title: DRAFTSMAN A

Titles now in use in this classification:

Layout Draftsman, Draftsman, General Draftsman

Minimum education and experience required:

Preferred—2-year vocational technical school that meets AIDD Certification requirements for "Draftsman" (See page 17), subject to a minimum of 1 year on-job experience in a temporary position at somewhat lower salary before fully qualifying for this position. ("Page 17" refers to AIDD manual.)

Alternate #1—3-year high school training which meets AIDD Certification requirements for "Junior Draftsman" plus 3 years drafting experience.

Alternate #2—High school graduate with 1 year of mechanical drawing, 1 year of algebra, 1 year of geometry, plus 6 years drafting experience.

Fair Labor Standards Act status:

Non-exempt

Supervision received:

Under general supervision with considerable opportunity for individual action.

Work direction exercised:

Might guide and instruct other draftsmen occasionally in this and lower classifications.

Supervision exercised:

None

Level of duties and responsibilities:

1. Handles normal drafting assignments under regular supervision.
2. Is completely familiar with drafting standards, symbols, nomenclature, engineering terms, proper use of materials, reference books and catalogs in a specific area of work.
3. Discusses job requirements directly with persons for whom work is being done.
4. Gathers information and data for jobs.
5. Makes routine calculations using standard engineering formulae.
6. Is assisted at times by other draftsmen in this and lower classifications; instructs, guides and checks their work.
7. Takes field or shop measurements as required.
8. Does limited amount of design under close supervision.

Figure 7 Sample Job Description

2.7 Cost Reduction and Performance Improvement Factors

Factors that decrease cost and improve performance are often closely related. Therefore, a new management or drafting technique that leads to reduced cost or improved performance of the D&D operation can be equally important from the manager's or company's viewpoint. They both add up to a stronger company and an improved competitive position.

A review of some sample techniques for increasing the D&D department's performance or reducing costs concludes this chapter.

CENTRALIZATION FOR GREATER EFFICIENCY

Centralization of the D&D effort provides numerous advantages in terms of improved performance and lower costs. First, it provides all D&D personnel with the leadership of the best available manager, allowing him to exercise his control directly and with a purview not readily available if he has his personnel scattered throughout the company and disconnected from his direct control. Consequently he can define and maintain more effective performance standards and take direct and positive actions to correct deviations from acceptable standards.

Centralization also facilitates uniform design and implementation of operating systems, closer coordination among supervisors, and improved communication throughout the department. Thus, problems can be detected earlier and positive corrective action taken immediately. In addition, management decisions can be made more quickly and with more information on hand than if several independent satellite groups existed spread through the company. (Note that even if satellite groups exist, the entire organization is still considered centralized as long as they are under the control of one manager.)

CASE IN POINT: Centralization pays good dividends when properly handled. For example, one company utilized its senior people in brainstorming sessions that saved the company over $60,000 in one year. The department manager used the following technique: He would solicit cost reduction recommendations from the staff in weekly or biweekly meetings. The staff consisting of two supervisors, an administrator, and a technical specialist would make recommendations or present recommendations of their subordinates. These recommendations were reviewed by all the key functional operations involved and a consensus on their value taken.

Selected techniques were then further analyzed to verify

their cost or quality effectivity and a report prepared summarizing the results. This report was critiqued by the staff and if no serious errors existed, the new technique or system was implemented.

The benefits of this approach towards cost reduction included elimination of duplicate forms, unnecessary prechecking operations, installation of new equipment, and reduction of unneeded checkprints. While these benefits could be achieved in a non-centralized operation, it is much less likely because of the broad scope of knowledge and coordination work that is required.

Of course, strong cases have been made for decentralized D&D operations. For example, one company has found that an increase in sales of about 23 percent required no increase in its engineering staff by using the project/team approach.[2] Thus, under certain conditions, cost savings may be available using decentralized organizational schemes.

ENHANCING COOPERATION, DEDICATION, AND PROFESSIONAL GROWTH

The individual—one of the most promising sources of cost reduction and improved performance—is often neglected. Although most managers would agree that the individual is the critical component in the man-machine-information triad, explicit and carefully developed efforts to cultivate this resource are rarely implemented. Therefore, a well-designed and consistently applied program for promoting a good psychological environment for greater productivity should be a prime element of an efficient D&D operation. A key aspect of this program should be a real interest in the staff members personal and professional welfare. Other important program elements include:

1. Maintain open lines of communication between management and staff.
2. Provide staff with timely information on company activities and progress.
3. Assign tasks to staff members which are challenging but neither overwhelming or too easy for their capabilities.
4. Offer fair pay rates proportional to creativity and productivity.
5. Provide a comfortable and pleasant working atmosphere.
6. Provide clear, positive, and unambiguous instructions for work assignments and responsibilities.
7. Provide up-to-date and reliable equipment, instruments, and furniture for effective accomplishment of work.
8. Provide periodic feedback from boss to employees and from employees to boss.
9. Provide clerical and support personnel to free D&D personnel from routine work.

10. Provide staff with opportunity to learn, grow, and earn more money.
11. Provide procedures and standards for all key operations.
12. Let staff know they are valued members of the company.

While the above rules seem simple to follow, their consistent and effective implementation is a difficult job. In fact, with time even an organization with effective personnel policies will find that interpersonal relations and company relations will subtly deteriorate until the preceding rules become just a list of nice sounding platitudes with no bearing to real-life policies.

Proper rewards for good people are essential for maintaining a dynamic group. Therefore, fixed salary limits (guidelines are okay) are not desirable where the output of the individual depends on his own motivation, skills, and knowledge. In addition, there is no valid reason in today's egalitarian society for the D&D manager to be the top paid person in the department. A highly skilled and productive designer should be rewarded equally if his contribution in his area results in comparable benefits to the company. In this way, exceptional D&D personnel can be retained in positions that would be filled by less competent people if higher pay can only be given by transfer to a management or other job. Besides losing a good designer, a bad manager may be obtained.

CASE IN POINT: One company had a rigid policy on raises within a given category but was liberal when someone was promoted into a new classification. As a result, when openings in design occurred, design draftsmen were eager to make the transfer. However, very often the design-draftsmen were not ready to become designers and became essentially high-paid draftsmen. The result was lower efficiency because people classified as designers could not perform in this category and senior draftsmen were promoted to design-draftsmen to fill the new openings created by the transfers.

To solve this problem, a new position could have been created to allow design-draftsmen to be recognized for special drafting abilities without requiring a change in their work requirements. Thus, they would have had additional status and money without being forced to take a designer's job before they were ready.

SPECIALIZATION FOR LOWER OPERATING COSTS

Specialization or labor differentiation can also save money by allowing lower paid personnel to be assigned jobs requiring less skill, and higher priced members with special talents to be assigned to work requiring their greater abilities. Thus, having a personnel mix to match the requirements of the company can be an efficient mode of operation under the right conditions.

Along with specialization is the more effective use of existing manpower. For example, it has been found through many studies that most companies used professionals at their maximum capabilities only 2 hours a day. The rest of the time they performed considerably lower level work.[3]

SHORT INTERVAL SCHEDULING FOR HIGHER THROUGHPUT

By breaking jobs into small subtasks, the output of the D&D department can be increased considerably.

CASE IN POINT: Using short interval scheduling (SIS), a manufacturing company found that it could cut its jobshop designers from 30 to 13 without a reduction in output. Its annual savings was estimated at $320,000.

Before SIS was started, the design section had a permanent staff of 25 designers and 30 job shoppers. This condition existed for 3 years until top management decided to attack the problem.

SIS was applied to this operation and within 7 months, the temporary staff was cut by more than 50%.[4]

Short interval scheduling involves breaking a task into short intervals by setting daily or weekly objectives for completion of definable work segments. The designer or draftsman is then required to report to his supervisor the status of each work segment per the established schedule. Experience has shown that a designer's output will be higher using this technique than if he is given a total task and asked only to report at the end of the estimated completion date.

The disadvantage of this system, however, is that it takes more work on the part of the supervisor to break down the job into small parts and to review progress each day with his draftsmen. Thus, if the supervisor cannot spend the time to follow-up the schedule on a regular basis, the results could be disappointing. Another problem that could arise is a negative reaction of the staff because of the tighter supervision and greater pressure applied.

AUTOMATED DRAFTING CAN LOWER LABOR COSTS

There are currently hundreds of automated drafting systems (ADS) in operation. The ADS is popular with large and medium-size companies because of the strong need to reduce their operating costs or to avoid excessive increases in D&D costs with growth.

CASE IN POINT: One year after implementing an automated drafting system, Raytheon's Equipment Division realized a 35 percent savings in generating all types of electro-mechanical drawings. A system consisting of digitizer terminals, one central processor, and a photoplotter was installed. A library of components commonly used by the company was also developed and inputted to the computer for rapid retrieval when needed and for ensuring standardization of drawings produced. The system was used to prepare mechanical details, schematics, logic diagrams, N/C tapes, and mechanical assembly drawings.[5]

At current ADS prices, a D&D department with as few as 15 draftsmen may profitably convert to an automated drafting system. With reduction in the price of this equipment, even small companies may find an ADS profitable.

PHYSICAL CHARACTERISTICS—Most components of an ADS are familiar to most managers, engineers, designers, and draftsmen. A typical ADS consists of drafting table, digitizer, keyboard, CRT input-display station, teletypewriter, magnet tape unit, and minicomputer. Computer output on microfilm units, card readers and punching units, and additional input-display stations are used with other equipment configurations.

The ADS can usually be housed in a separate room about 300 feet square. Because of the intricate nature of the equipment, it is best to keep it in a controlled-access room so that people won't play with or accidentally damage it. Although the system operates in normal lighting, the ability to dim lights is desirable for operator comfort.

A typical ADS drafting table has a 42 by 60 inch working surface. Larger tables are available.

ADS's work off normal electrical power, e.g., 110 vac, 50 amperes of single phase power. However, current requirements can exceed 80 amperes, depending on how much equipment is purchased. Separate power lines are usually recommended for minimizing noise that can cause the system to malfunction.

FUNCTIONAL ELEMENTS—An ADS system consists of three basic functional elements: input, processing, and output. A review of the various subelements within each of these categories follows:

Input

Digitizer: Converts location of lines, points, and components on rough sketch into X-Y coordinates for computer processing. The operator positions a cursor over the point to be digitized and presses a button to transfer the coordinates to the processor.

CRT Display: Provides visual display of information just inputted or previously stored in processor memory. Allows presentation of only part or all of the drawing of interest. Allows real-time man-machine interaction; e.g., corrections, additions, and changes in component locations, lines, and notes.

Menu: A collection of standard operations, commands, and symbols commonly used in the preparation of drawings by the user.

Alphanumeric Keyboard: Allows operator to add letters, words, paragraphs, and numbers to drawings when needed.

Function Keyboard: Allows operator to tell processor what he wants to do, e.g., erase, move a line or symbol, repeat a symbol, and add a grid or a rule.

Sensing Pen/Stylus: Used to transfer input data from a sketch onto a CRT display or into processor. Also can be used to draw on the face of a CRT, with the processor automatically producing straight lines, providing proper spacing, making perfect circles or curves, or detail components. All systems using CRT's do not have the capability of drawing directly on the CRT face. They may use tablets with a sensing pen for inputting data onto CRT.

Card/Tape Reader: Converts data recorded on punched computer cards, paper tape, or magnetic tape into digital data fed into processor.

Processor/Computer

Processor: The processor consists of a control unit; central processing unit; various memory devices for storing library symbols, drawings, and computer programs; and specialized software developed for allowing the system to produce finished drawings from digitized input data. (Note: a library of symbols consists of special sets of symbols commonly used by the company on its drawings.)

In addition to basic control functions of tape search and read, data buffering, and velocity/acceleration control of drafting head, the processor allows operator control of data scaling, origin selection, pen selection, and image rotation.

Storage Devices: Memories include disk, tape, cassette, and core. Cores are used for processing data, disks and cassettes for temporary/permanent drawing storage, and tape for mass storage of the many drawings produced by the company.

Teletypewriter: A teletypewriter is also used to supply operator commands to the computer and to output messages such as incorrect operations, to him.

Output

Plotter: The prime output device is the plotter, which converts data fed from the processor into the finished drawing. Different types of pens can be used to produce various line widths. Plotters consist of flatbed and drum types.

Photoplotter: A photoplotter uses a focussed beam of light to prepare artwork by exposing sensitized film to the light. The photo-exposure head is moved to correct locations and then the beam is turned on and off under the guidance of the processor to prepare very accurate artwork patterns for printed circuit boards.

Hard Copy Printer: A hard copy printer is sometimes used for producing fast copies or checkprints, without tying up the slower, more expensive plotter.

Teletypewriter: Used for outputting textual information such as messages, parts lists, and X-Y coordinates.

COM: Computer output on microfilm (COM) can be obtained with special equipment. In this case, the output could be aperture cards, microfiche, or roll microfilm.

Magnetic Tape Recorder: The computer can output to a magnetic tape unit for mass storage of drawings.

Paper Tape Punch: A punch can be driven by the computer to produce paper tape for driving plotters, artwork generators, or drill machines.

OPERATION—To prepare a drawing using an ADS, the operator tapes or attaches a rough sketch prepared by an engineer or designer to the digitizer table. He then follows these steps:

1. He points with a free moving cursor to the desired operating instruction on the menu (e.g., add, move, or change).
2. He then moves cursor over the picture of a desired symbol on menu.
3. He points cursor to place on drawing where he wants the symbols to appear and presses a digitize button to record the symbol at the desired location.

The preceding procedure is repeated for various symbols, lines, and circles until the drawing is completed; i.e., all information on rough sketch has been inputted and stored in computer memory. The keyboard is used for adding alphanumerical information such as notes and dimensions to the drawing and is stored in the computer memory with the graphical data.

While the drawing is being created by the draftsman or designer, it can be checked for accuracy by viewing its digitized portions on the CRT. Correction or changes (editing) can be made immediately. Note that portions or all of the information inputted to the computer can be displayed on the CRT. When the drawing is complete and correct it can be plotted on paper, vellum, or film while the operator is digitizing another sketch. However, new drawings need not be plotted immediately. The information stored in the computer's memory can be outputted to magnetic tape or other outside storage medium for off-line plotting or storage when desired.

Note that the preceding operations vary from system to system; therefore, the procedure will depend on the equipment purchased and could be considerably different.

CAPABILITIES—Automated drafting systems have an exceptionally broad range of capabilities. However, each system has its own design characteristics and capabilities and limitations, depending primarily on its software. Therefore, based on the ADS selected, a wide range of capabilities are currently available in the area of drafting, e.g., schematics, assemblies, maps, flow charts, IC masks, perspectives, PERT charts, and mechanical details.

An experienced operator can produce more than 400 drawings per year using an ADS, including a down time for maintenance and repairs of 5 percent for 24-hour operation.

TRAINING PROGRAMS FOR IMPROVED PERFORMANCE

A training program that is implemented on a consistent basis can increase the operating efficiency of the entire department. Training programs, however, should not become routine, innocuous efforts that lack dynamic participation of the staff.

In addition, training programs should include senior staff members. Even experienced designers and draftsmen forget things and can benefit from reviews of standard procedures and techniques. Also, as new methods and procedures are established in the field, they must be spread to all who can benefit. An example of how one company derived benefits from an in-house training program follows:

CASE IN POINT: One medium size company decided to give seminars on true positioning because ignorance on the part of inspectors, designers, and engineers was creating unnecessarily scrapped parts, arguments on whether a part was good, or what a drawing meant.

A post-seminar survey of the participants indicated that they had a much better understanding of what true positioning and dimensioning was about. In addition to its educational value, the staff found a faster way of inspecting parts using a special template developed by the instructor. This template was used to inspect typical parts, and the time to complete this work compared to the traditional method of inspection was reduced sharply. The annual savings resulting from the use of this template was $7,500.

RAPID DATA RETRIEVAL SAVES MANPOWER

Rapid and convenient data retrieval can help expedite D&D work. Considerable portions of new work can be done using old drawings, specifications, and parts lists. Therefore, if a designer or draftsman has to spend an hour or more looking for an old drawing and getting it copied for use in preparing a new document, cost and time requirements will increase.

> **CASE IN POINT: Lovelace-Lawrence and Co. reported that a company using its centralized engineering data bank had remarkable results in retrieving design data quickly. The company had set up a data bank with Lovelace-Lawrence consisting of 200,000 items for design retrieval. The company's engineers and designers found that nearly 50 percent of their searches of previously released designs on other projects resulted in usable designs using this data retrieval system. About 25 percent of these retrievals were directly usable and the other 25 percent required minor modifications.**
>
> **The system permits the designer to describe the type of part or class of design needed and gives him reproductions of all the designs in the file which meet his requirements. Retrieval is achieved in seconds using this rapid access data retrieval system.[6]**

The solution for slow data retrieval depends on the size of the company. For a small one, assignment of responsibility for maintaining orderly and complete engineering files to a reliable person can do the job. For larger companies, with many thousands of different documents, an automatic data retrieval system may be needed using microform e.g., aperture cards, microfilm, or microfiche. Once the desired reduced-size document is retrieved, the draftsman may take a duplicate microform to his work table and insert it into a desk type microform viewer, which allows him to use the old document as a guide for preparing the new one. A sample system is shown in Figure 8.

CURRENT AWARENESS PROGRAM: INFORMATION FOR HIGHER EFFICIENCY

Being aware of the latest technological developments related to company activities can promote improved performance and lower costs. Several tech-

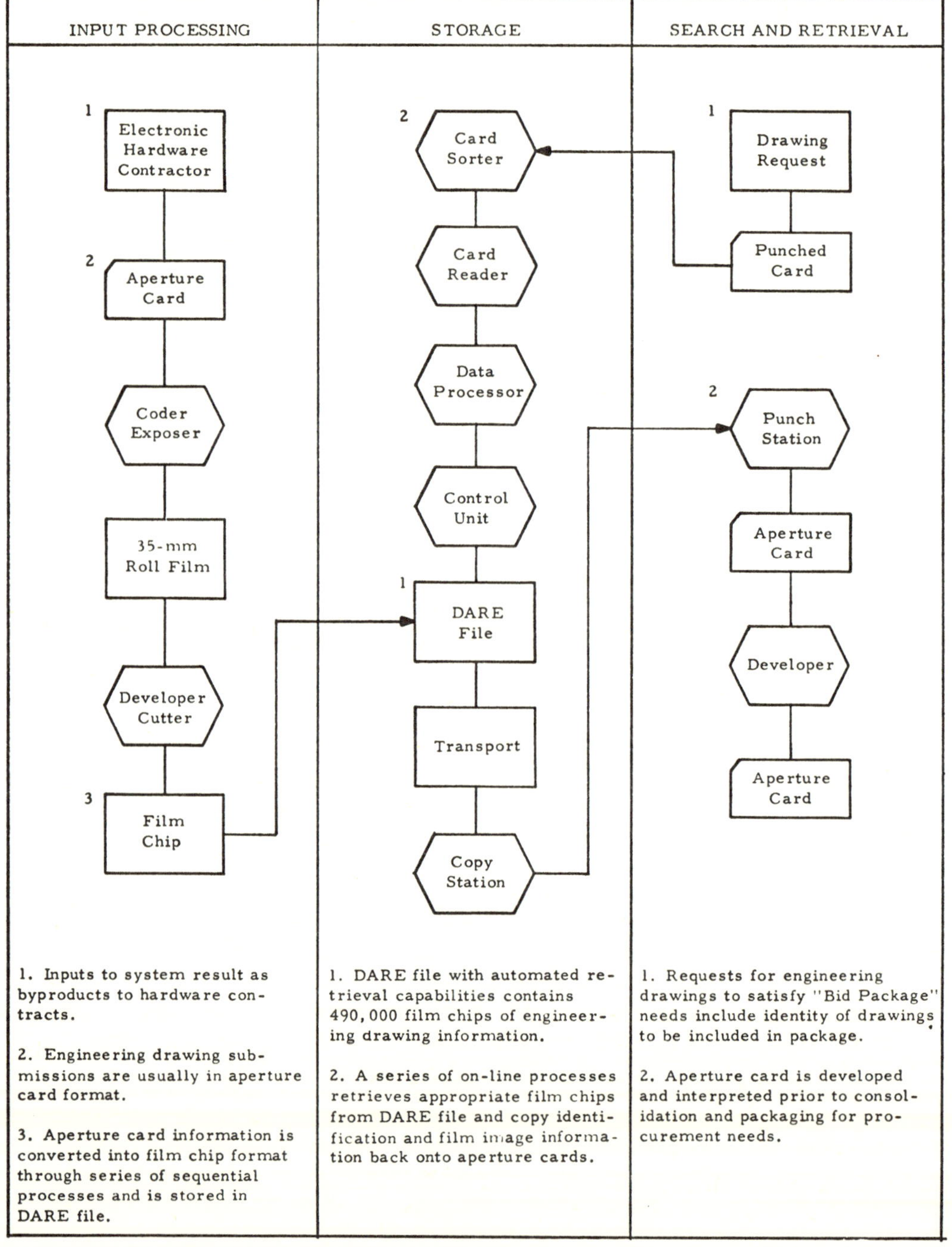

Figure 8 Sample Data Retrieval System

Source: "Information Retrieval Systems," General Services Administration, US Government, FPMR 101-11.3, 1970, p. 33

niques can be used to obtain new information related to D&D and company activities. These include:

1. *On-site Information Center*—Information analysts review new literature (journals, books, reports, and papers) for information of interest and send abstracts of the documents to D&D for use.
2. *Trade Journals*—D&D personnel review trade and professional journals in fields of interest.
3. *Personal Contacts*—Personal contacts are maintained with people doing advanced work in the field.
4. *Seminars and Conferences*—Staff members can attend professional seminars/conferences periodically to acquire new information.
5. *Abstract Services*—Subscribe to abstract services for convenient review of articles or books of interest.

An example of an information service called SCIM is presented next to illustrate the scope and method of obtaining information on new developments. SCIM stands for "selected categories in microfiche." It is an information source of the National Technical Information Service (NTIS) of the US government. SCIM is a standing order service that sends microfiche copies of new reports in fields of special interest to subscribers on a semimonthly basis. A 4 by 6 inch sheet of film with up to 98 images arranged in rows is sent whenever a report in a selected category is published. An inexpensive (about $150) portable microfiche viewer can be used to read the small documents on this film. Printers are available for making full size pages from the film when needed.

Subscription to this service can be by agency (e.g., NASA, Department of Commerce, Department of Defense, or Atomic Energy Commission) or by category. Thirty-nine categories are available such as:

—Reprography and record devices
—Building technology
—Library and information sciences
—Management practice and research
—Space technology
—Industrial and mechanical engineering
—Civil, structural, and marine engineering
—Computers, control theory, and information theory
—Electrotechnology
—Environmental pollution and control

This service is provided by NTIS, SCIM Division, National Technical Information Service, US Department of Commerce, Springfield, Virginia 22151.

OTHER SOURCES OF INFORMATION—The Defense Documentation Center, Naval Forms Center, National Aeronautics and Space Administration, and trade associates can be sources of valuable information for D&D personnel. For their information services and documents, write to:

Defense Documentation Center
Attn: DDC-TAR
Cameron Station, Alexandria, Virginia 22314

Commanding Officer (code 512)
Naval Publications and Forms Center
5801 Tabor Avenue, Philadelphia, Pa. 19120

NASA Scientific and Technical Information Facility
Post Office Box 33, College Park, Maryland, 20740

National Standards Association, Inc.
1321 14 Street N.W., Washington, D.C. 20005

Aerospace Industrial Association of America
1725 DeSales Street, N.W. Washington, D.C. 20036

American Institute of Architects
1735 New York Ave., N.W., Washington, D.C. 20006

American Welding Society
2501 N.W. 7th Street, Miami, Florida 33125

Society of Automotive Engineers
2 Pennsylvania Plaza, New York, N.Y. 10001

American Society of Mechanical Engineers
345 E. 47th Street, New York, N.Y. 10017

American Society for Testing & Materials
1916 Race Street, Philadelphia, Pa. 19003

American National Standards Institute (ANSI)
1430 Broadway, New York, N.Y. 10018
Note: ANSI includes former American Standards Association (ASA) and US of America Standards Institute (USASI)

National Aerospace Standards Committee
Aerospace Industry Association of American, Inc.
1725 DeSales Street, N.W. Washington, D.C. 20036

American Ordnance Association (American Defense Preparedness Association)
819 Union Trust Bldg. Washington, D.C. 20005

Construction Specifications Institute
1150 17 Street N.W., Washington, D.C. 20036

Many trade associations exist for specialized disciplines. If information on a particular type of association is desired, write to the following for help:

Director of National Trade Associations
Department of Commerce, Washington, D.C.

Libraries also have reference books on trade associations. See *Encyclopedia of Associations, Volume 1, National Organizations of the US.*

If current information on standards and standardization activities is needed, contact:

Standards Information Services, National Bureau of Standards, Washington, D.C.

Current information on new D&D practices and activities can be obtained from various trade journals. A few of these are listed below:

Reprographics & Engineering Records Management (formerly *Graphic Science*)
750 Third Avenue, New York, N.Y. 10017

Engineering Graphics
25 W. 45th Street, New York, N.Y. 10036

Design & Drafting News
American Institute of Design and Drafting
3119 Price Road, Bartlesville, Okla. 74003

Plan and Print
10116 Franklin Ave., Franklin Park, Ill. 60131

Business Graphics
7373 N. Lincoln Avenue, Chicago, Ill. 60646

Other journals include *EDN, Electronic Design, Electronics, Machine Design, Process Engineering, Engineering Design Graphics Journal, Product Engineering, Reproduction Engineering, Standards Engineering, Ordnance*, and many other trade journals in specific scientific and engineering fields.

Additional information can be obtained from the following organizations:

AMERICAN INSTITUTE OF DESIGN AND DRAFTING (AIDD)—AIDD is an organization dedicated to improving the science of graphics communication. It provides leadership in the D&D field by conducting conferences, collecting data on the field, and publishing needed works such as drafting manual guides, job description standards, and other documents. It also certifies design-drafting curricula in schools.

An organization of this type offers up-to-date information on new developments in D&D that can lead to cost reductions and improved operational performance. Thus, members can obtain information on management skills, drafting materials and equipment, and general drafting room operation.

Additional information on the activities of AIDD can be obtained from: AIDD, 3119 Price Road, Bartlesville, Okla. 74003.

AMERICAN SOCIETY OF PROFESSIONAL DRAFTSMEN AND ARTISTS—This is another organization that is dedicated to graphics communication improvement. Information on its goals and activities can be obtained from: American Society of Professional Draftsmen and Artists, 415 St. Paul Place, Baltimore, Maryland 21202.

INTERNATIONAL ASSOCIATION OF BLUE PRINT AND ALLIED INDUSTRIES—For information on this organization's activities write to: International Association of Blue Print and Allied Industries, 10116 Franklin Avenue, Franklin Place, Illinois 60131.

Note: The addresses of publications and organizations change over the years. Therefore consult the *Industrial Arts Index* for journals and *Encyclopedia of Associations* for trade organizations. Both of these reference books are at your library.

SEMINARS/CONFERENCES—Seminars and conferences offer opportunities to find out about new drafting and reproduction equipment, techniques, standards, management procedures, and technical developments. Although the presentations can provide more information than most people can absorb in a few days, collection of the speaker's papers can allow most attendees to selectively study the subjects covered so that they can be applied to their work.

Although performance improvement and cost savings techniques can be obtained from these seminars and conferences, they are usually oriented towards large D&D operations. Small companies, therefore, have to be especially careful about automatically assuming that they can profitably apply the techniques presented. Careful analysis is required before adapting the sophisticated methods of operation used by larger companies.

STANDARDIZATION REDUCES COSTS AND DELAYS

Developing standard operating procedures, job descriptions, and formats has proven highly cost effective in industry. For example, the use of standards can save thousands of dollars per year, even for small companies. The American Standards Association conducted a study in the savings obtained from standardization programs that revealed interesting results. It found that for every dollar the rocket industry invested in standardization, it received $415 in cost savings. This represented the greatest savings of any type of industry. But even the lowest return was 35 percent more than invested.[7]

Standardization procedures should be applied whenever routine patterns, materials, or activities exist.

CHECKING REDUCES WASTED MATERIAL AND LABOR

Although most large companies use professional checkers to minimize errors and to increase the quality of their drawings, many medium and small companies often try to live without these people. Experience shows that an effective checker can save a company many times his annual salary by catching errors that would have required scrapping, extensive rework, and delays after the product was built. He can also save money by suggesting better or cheaper methods of manufacturing or designing an item.

Note that a self-check by the draftsman often leads to repeating the same errors, and a check by his supervisor may be a cursory one if the supervisor is busy with administrative tasks requiring immediate attention.

As a cost saver, the checker should be a responsible individual who will not allow personal prejudices and preferences affect his changes to the drawing. The overriding criterion should be no change if the drawing is adequate for its purpose and the customer's needs.

When a checker is used, colored pencil markings should be used to quickly and positively identify the type of error found. For example, one system requires marking every item on a drawing, including titles and notes, with colored pencil. The following color markings are used:

YELLOW—indicates that the item shown is correct.

RED—indicates that a mandatory change is required. Extensive changes may be indicated by a note stating: "Revise per document X, Sec. 2, p. 25."

GREEN—indicates that the item marked is questionable and needs the attention of the engineer.

To aid the small company in this cost saving function, a checklist of drawing items that should be reviewed by a checker is given in Table V.

Table V
23 Point Drawing Checklist

1. Acceptability of design approach
2. Conformance to engineer's or designer's layout drawing
3. Impractical manufacturing requirements
4. Correct material callout
5. Acceptability and accuracy of parts and materials identification
6. Strength and rigidity of structural design
7. Economy of design and parts selection
8. Clearance and tolerance acceptability
9. Identification and marking of item requirements
10. Specification of quality assurance and test requirements
11. Accuracy and completeness of line weights, arrowheads, spelling, scale, abbreviations, and title
12. Drawing number callout
13. Data in title block adequately filled out
14. Next assembly and used on (end item) data
15. Parts list and call out agreement

16. Adequacy of notes
17. Approval signature completeness
18. Security classification
19. Adequacy of views, sections, and dimensions
20. Specification of weight of item
21. Use of standard parts when possible
22. Accessibility for maintenance and repair
23. Proprietary/confidential classification

VARIABLE QUALITY LEVELS FOR IMPROVED EFFICIENCY

By identifying and defining different quality levels for engineering data, considerable savings can be obtained. For example, D&D may recognize four distinct quality levels that meet different needs of the company. These could range from top quality drawings to free-hand sketches.

Unnecessary preparation of documents to a top quality level can increase D&D costs by many thousands of dollars per year. Therefore, once the categories are defined and their areas of acceptable usage identified, they should be called out by the drafting supervisor on the drafting work request.

REPRODUCTION TECHNIQUES CAN CUT COSTS

Reproduction techniques offer many ways of reducing costs. Table model whiteprinters in the immediate areas of the D&D group can reduce walking and waiting time for checkprints. Folding machines can reduce clerical labor by eliminating manual folding operations and reducing machines can lower paper, storage, and handling costs.

> **CASE IN POINT: Sperry Gyroscope reduced cost of manual folding of prints by mechanizing the operation. A Printfold folding machine was installed with a Bruning blueline printer. The prints were reproduced and stacked on a requisition basis in quantities ranging from 3 to 300 prints for each order. As the orders were completed, the prints were folded and prepared for distribution. By introducing two automatic folding machines into its operation, the company saved $18,000 a year in reduced labor.[8]**

Chapter 8 describes other techniques for reducing reproduction costs.

STANDARD PARTS CATALOGS AND REFERENCE DOCUMENTS SAVE MANPOWER

Standard parts catalogs can save designers much time in looking up the characteristics of items specified by an engineer or selected by the designer himself. Services are available for providing catalog or specification sheets to cover thousands of parts. These services also automatically supply new or revised catalog or specification sheets when they are released by manufacturers.

One parts information system contains data on microfilm which is displayed on a viewer. If the information displayed is desired, the operator pushes a button to obtain a full-size copy for his own use.

Reference documents are standards, specifications, and tables of technical data useful to D&D personnel. The types and number of documents available depend on the company's business. When readily available, they can help the designer or draftsman complete his work more swiftly and reduce errors. Microfilm systems are available to speed up retrieval and revision of military specifications, handbooks, standards, and qualified products lists.

When Government documents such as specifications and standards, are required immediately for completing design work, document supply companies can be contacted for delivery of the desired documents at about 10¢ a sheet. These companies can deliver the documents ordered within 1 or 2 days.

TEMPORARY HELP

Periodic fluctuations in manpower needs are common to many companies today. Manpower needs may increase for only a few days or for many months. However, to avoid the disadvantages of employing new D&D personnel for a short time and laying them off, many companies use temporary help agencies to supply manpower during peak work loads. The disadvantages of hiring and firing involve more than higher personnel acquisition costs and can include lower morale and a bad company reputation. These in turn lower overall productivity and discourage competent personnel from seeking employment with the company.

The advantages of using temporary help include:

- More rapid acquisition of needed personnel than if normal employee requisition, advertising, and interview operations are followed.
- Lower acquisition costs because the temporary help agency distributes its costs over all its customers. In addition, because it specializes in this area

and has a large reservoir of designers or draftsmen to call upon when needed, the acquisition costs are smaller. The agency's lower overhead allows more savings to be passed on to its customers.

- The cost of temporary employees can be less than the burdened cost of regular employees; e.g., a temporary draftsman may cost $7/hour versus a burdened cost of $12/hour for a regular draftsman.

CASE IN POINT: One small company saved $1486 by using three job shoppers for a short term peak work load. These job shoppers were obtained from a local job shop representative who found the draftsmen with the qualifications needed by the company, sent them down for an interview with samples of their work, and sent down another candidate when one of the draftsmen didn't meet the company's requirements. Because of the skill level and lower burdened cost, the work was completed before schedule and at a total lower cost. In addition, recruitment costs were not incurred.

- Overtime can be reduced or eliminated thereby reducing total costs due to higher efficiency and lower hourly wages (time and half can be eliminated).
- Means to obtain permanent people after a trial period.

Of course, the use of temporary help has its pitfalls, especially when an agency is used for the first time. In this case, the personnel supplied may not be what are needed and their output may not be as high as desired. However, if some experience is built up gradually, when a major increase in staff is needed, the procedures for defining requirements, preparing facilities for the temporary employees, and coordinating other related activities will be much more efficient. Another pitfall is an inexperienced supervisor who does not know how to plan for the efficient use of temporary employees. As a result, facilities, work, and direction are not available when needed and the temporary help sits around with nothing to do.

Information on job shoppers can be obtained from the National Technical Services Association: NTSA, Suite 315, 1225 Connecticut Ave., N.W. Washington, DC 20036.

STUDENT HELP—Students from local technical schools and colleges can also be used to reduce overtime work and to perform simple, routine drafting work. When conditions are right, students can be valuable assets to the D&D department and can provide the company with a basis for determining whether a student should be hired after graduation.

FACSIMILE EQUIPMENT FOR INSTANTANEOUS TRANSMISSION OF DATA

Facsimile equipment can be used to rapidly transmit drawings or parts of drawings to distant locations. Although restricted to small size sheets, this equipment allows portions of a product configuration to be sent to another area where they can be taped over the part of the drawing that has been changed. Thus, service or other groups get prints of changes as they are released.

AUTOMATED WORD PROCESSING

Automated word processing using a computer can reduce costs when frequent revisions or print outs of documents are required. For example, standard operating procedures often require minor changes to accommodate changes in operations. An automated processing system avoids having to retype the entire procedure each time it is changed. The system can also be used to prepare parts and data lists, manuals, procedures, and specifications. These documents are stored in a computer or off-line memory components. A typewriter terminal is used to input and output documents. High-speed printers are also used for outputting final documents.

Note that this and other sophisticated techniques don't always work, especially if they are not implemented systematically.

CASE IN POINT: Based on the experience of one of its staff at another company, a medium-size company installed an automated word processing system. The system was placed under the direction of the Drafting Manager. After 1 year of operation, it was found that the system was not being fully utilized and the cost savings anticipated not being realized. The system was transferred to the data processing group which also failed to use it profitably.

While the potential benefits were real, the system failed because of the following reasons:

- **The user's interest in using the system was not determined or solicited to assure their cooperation.**
- **A single, interested person was not assigned to coordinate and implement the system.**
- **Systematic evaluation and follow-up activities were not implemented to allow management to be informed as to the efficacy of the word processing system.**

TEAM DRAFTING

Team drafting involves a number of people working on one drawing. The system lowers costs through better utilization of skill levels. An experienced draftsman acts as team leader and directs semi-skilled people working on the routine aspects of the drawing. The draftsman is thus allowed to concentrate on the more technical or creative aspects of the drawing. The team usually consists of the following:

- Draftsman: provides technical direction and coordination. May supervise several teams.
- Viewmaker: produces simple views.
- Dimensioner: aligns parts on standard format and adds dimension and support data.
- Clerk-typist: types notes and other information normally lettered by draftsman.

DRAWING RESTORATION

Restoration of old drawings instead of redrawing them can save many hours of labor. The need for restoration occurs when drawings have had considerable usage, have been damaged or defaced, or have a new format requirement to include a change in the company's name or address.

The use of a xerographic printer allows the restoration process to be simplified and requires someone less skilled than a draftsman to do. The procedure is as follows:

1. Make a copy of drawing so that all the information is legible. This may require running it dark and some background will be evident.
2. Darken any light lines or characters.
3. Use white china marker, artist's white tempera, or white correction fluid to cover any excessively dark background area.
4. Place film positive drawing format on top of print if used.
5. Run full-size print on vellum paper using special xerographic equipment.

OTHER TIME-SAVING TECHNIQUES

Various techniques exist for reducing drafting costs. For example, drawing vellums can be revised without completely redrawing them by cutting and pasting. Altered parts are cut out and new ones pasted or taped to the old vellum. When xerographic equipment is available, the new section (on bond paper) can be taped over the obsolete area of an old print, which is reidentified and used as a new original.

When extensive erasures of reproducibles (brownlines or sepias) are required, special film reproducibles are available that can save much erasing time. Thus a mildly damp eraser can easily remove images or letters quickly and without special chemicals. An electric eraser can also save a lot of time.

Other techniques for reducing drafting costs are described in Chapter 4.

3

3.0 Data Control Facility: Key to Document Integrity and Protection

The data control facility is responsible for issuing approved documents (data) to various departments within the company or to a customer. Other equally important functions are the prevention of unauthorized changes to released documents, the storage and retrieval of documents, and the protection of these documents from loss or damage.

Along with these functions, the data control facility may reproduce data for distribution to the staff if some other department is not responsible for this operation.

Accurate labeling or marking of documents with usage data is also a critical responsibility. This involves identifying each document in terms of its acceptability for use by production or purchasing. When documents are not to be used for these purposes, other designators are added to the documents such as reference, inspection, or checkprint.

BENEFITS AND APPLICABILITY OF THIS CHAPTER

The principal benefit of this chapter is in highlighting the importance and techniques of protecting property equivalent in value to precious metals. The common drawing, worth about $200 per ounce, is an extremely valuable part of the company's resources; therefore, its protection is necessary. Specific benefits of this chapter are:

1. Gives review of facilities and equipment needed to control and protect data.
2. Provides a checklist for a data release system.

3. Describes staff requirements for an effective operation.
4. Presents sample records, forms, and special files used in data control activities.
5. Provides a list of retention periods for common engineering data.
6. Provides descriptions of various types of releases and their purposes.
7. Provides cost reduction techniques.

The techniques covered in this chapter are, for the most part, necessary for all companies. Of course, the scope and sophistication of facilities and equipment and staff have to be tailored to the size of the company. As mentioned, if the applicability of a technique or system doesn't seem practical, use Table II to evaluate it.

KEY OBJECTIVES

Key objectives of the data control facility are listed below:

- Rapid processing and distribution of released documents.
- Complete physical protection and security of documents.

Note that to meet these objectives, the data control facility needs a high level of autonomy and independence from other departments. Therefore, it should not be under the direct control of the D&D department. Instead, it should report to an engineering services manager, director of engineering, administrative manager, or general manager.

What kinds of documents should be kept by data control? Ideally the data control facility should keep every official document released by the staff. Thus, an integrated document center under the control of one supervisor provides a highly organized and efficient operation that offers professional services for all members of the company. Unfortunately this is not always feasible for several reasons. These include: traditional patterns of operation are hard to change, control of accounting and computer systems is strongly entrenched in a different department, and trained data managers are scarce.

However, regardless of what the rest of the company does with its documents, D&D, engineering, R&D, production, testing, quality assurance, facility, and technical service groups should submit their document originals to data control for safekeeping. It is also very important to control changes to equipment installation and facilities construction documents. With increasing emphasis and Government involvement with safety, documents related to this area must also be controlled.

The data control and reproduction cycle is shown in Figure 9. Note that

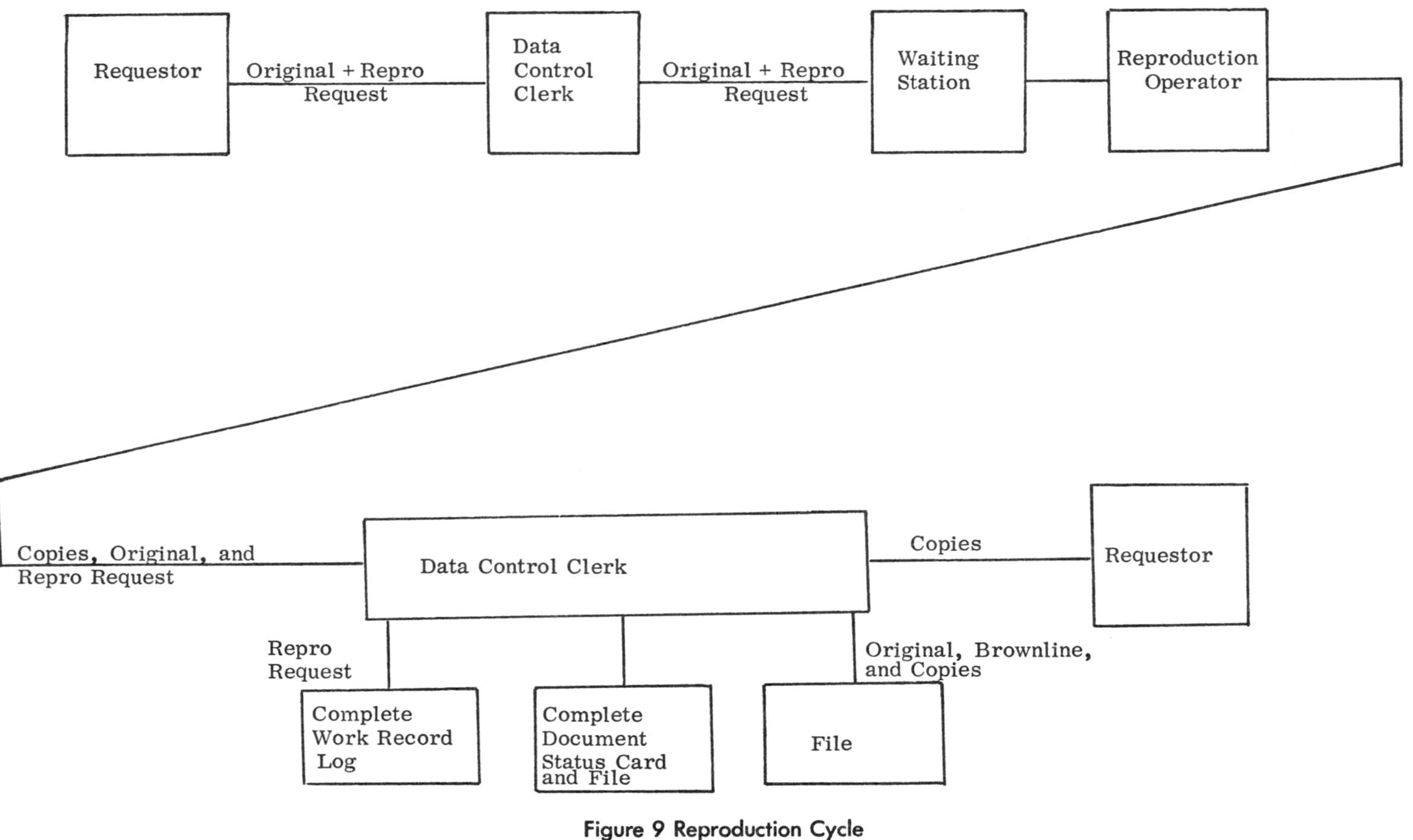

Figure 9 Reproduction Cycle

regardless of company size, the basic system must exist, if only informally, to prevent chaos in the documentation area.

CASE IN POINT: It was the practice of one small company to allow control of drawing originals to be done on an individual basis. Shortly after a draftsman left the company, it was found that four drawing originals had also disappeared. These originals may have been thrown out when he left or he may have taken them with him. However, since the draftsman had taken a job out of state, it was difficult to trace him. As a result, the cost to redraw them was about $800. In addition to this kind of loss, drawings handled in this way are often misplaced or subject to faster wear and can cost many hours per year trying to find or repair them.

A major function of data control is to provide information on the configuration of parts, equipment, and facilities. A checklist of what information should be retrievable from records and data stored in the facility is given in Table VI.

Table VI
Checklist for a Data Release System

- Can the part numbers used in a particular component, sub-assembly, or assembly be found?
- Can the part composition (part and serial numbers) of each product built be determined?
- Can the next higher assembly drawing and part numbers for a particular item be found?
- Can the end item in which a part is used be determined?
- Can the change identification documents released for each item be located?
- Can the effectivity of each change document be found?
- Can the specification control documents for parts built by vendors/subcontractors be identified?
- Can the standard specification numbers and standard part numbers for parts used within any non-standard item be determined?
- Can the end item's specification number be found?
- Can the manufacturer's code identification number for all vendor and subcontractor-supplied parts be determined?
- Can superseded and obsolete drawings and parts lists prepared during the last 2 years be found?
- Can the people who received copies of a particular document be identified?
- Can the person who checked out a document original be identified?
- Can the person who authorized the release of a document be identified?
- Can a record of a field retrofit be found?

3.1 Facilities and Equipment: Essentials for Control

The facilities of data control include filing cabinets for storing released data, e.g., standard vertical files, legal size files, flat files, open shelf files, and automated files.

Besides document files, record files are needed such as rotary desk, tub, card, and notebook files. The quantity and sophistication of these files depend on the quality and response time requirements imposed on the data control facility.

In addition to the preceding, work tables, desks, chairs, storage cabinets, rubber stamps, binders, paper cutters, and paper shredders are needed. If the facility includes reproduction equipment, white printers, office copiers, microfilm camera and duplicators, collators, folding machines, industrial photographic equipment, and contact printers are necessary.

A key aspect of the facility should be a closed access. That is, no through traffic by the staff should be possible. This can be achieved by using a dutch door, i.e., top half opens up for transacting business with the staff while the bottom half remains closed.

Another aspect of the facility should be structural integrity. Sometimes fireproofing is needed to ensure protection of documents. Of course, a fire extinguisher should also be near by. Additional protection can be provided by special cabinets that have thick insulation for better fire resistance.

In some cases, records may be stored in a commercial warehouse, a bank's safety deposit vault, or an underground secruity area.

Good locks, strong walls, and alarm systems can also help reduce industrial espionage. While few company's implement all these security features, they should be carefully evaluated against the probability of losing documents and the consequences on company operations.

3.2 Staff: Special Disciplines For Effective Data Control

The staff of a combination data control and reproduction facility includes a supervisor, data control clerks, photographer, and reproduction machine operators. The number of clerks and machine operators, of course, depends on the size of the company and its documentation requirements. For a small company, a supervisor and clerk-reproduction operator may be adequate for most of the time; temporary help is used during peak workloads. For larger companies, several clerks and machine operators may be needed.

SUPERVISOR

The supervisor is responsible for organizing the facility and making sure that policies and procedures defined by management are being implemented. In

Engineering Release and Issue

Title	Prefix	Document Number	Change Letter

Job Number	Configuration Item/Model Number/Type	EID	Next Higher Assembly/Document

Date of Release by Engineering	Anticipated Release Date	Required Release Date	Originator	Department Number

Size	Level or Grade	Distribution	Number of Sheets		Security

☐ Formal ☐ Drawing
☐ Record Only ☐ Specification
☐ Original ☐ Procedure
☐ Change ☐ Parts/Wire/Drawing List
☐ EO, ADCN, DCN ☐ Schematic
☐

Remarks and Limitations

Effectivity		CCB Directive	Release Authorized by	Date	Department Number
From	Through				

Type of Release	Issued by	Issue Date
Production ☐ Reference ☐ Limited ☐ Advance		

Figure 10 Engineering Release Form

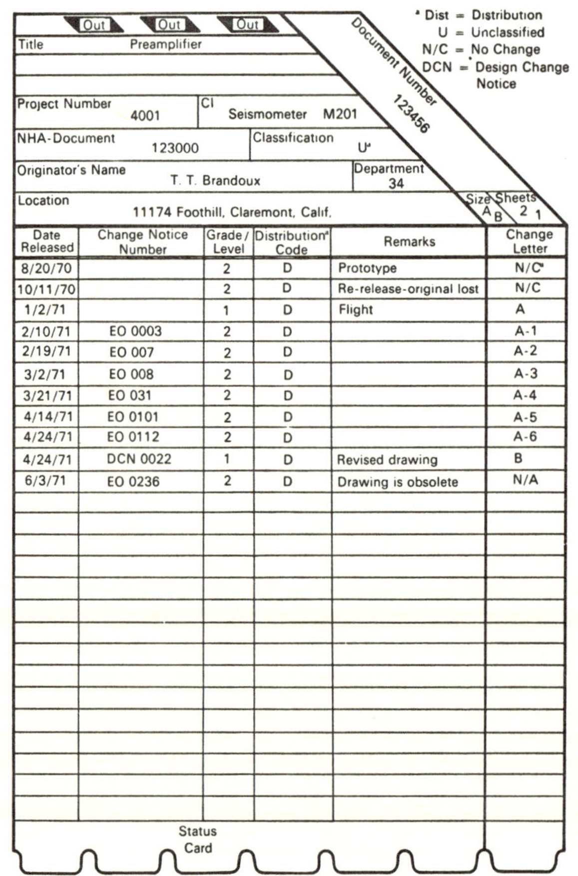

Out Out Out

Title Preamplifier

Document Number 123456

Project Number 4001 | CI Seismometer M201

NHA-Document 123000 | Classification U*

Originator's Name T. T. Brandoux | Department 34

Location 11174 Foothill, Claremont, Calif. | Size A B | Sheets 2 1

Date Released	Change Notice Number	Grade/Level	Distribution* Code	Remarks	Change Letter
8/20/70		2	D	Prototype	N/C*
10/11/70		2	D	Re-release-original lost	N/C
1/2/71		1	D	Flight	A
2/10/71	EO 0003	2	D		A-1
2/19/71	EO 007	2	D		A-2
3/2/71	EO 008	2	D		A-3
3/21/71	EO 031	2	D		A-4
4/14/71	EO 0101	2	D		A-5
4/24/71	EO 0112	2	D		A-6
4/24/71	DCN 0022	1	D	Revised drawing	B
6/3/71	EO 0236	2	D	Drawing is obsolete	N/A

Status Card

* Dist = Distribution
U = Unclassified
N/C = No Change
DCN = Design Change Notice

Courtesy Wiley & Sons, Samaras and Czerwinski, *Fundamentals of Configuration Management*, Wiley-Interscience, 1971.

Figure 11 Document Status Record

addition, he monitors the facility's operation to detect possible problem areas in document security, lowered production capabilities, and personnel morale and conflicts. He is also constantly looking for improved methods of operation and new equipment that can reduce costs and improve efficiency.

The supervisor is also responsible for knowing the current contents of the data control manual, if one exists, and for instructing new personnel in its procedures. When changes are made to procedures or equipment, he is responsible for training the staff in their use.

DATA CONTROL CLERK

The data control clerk is the key interface point between the facility and the rest of the company. The clerk performs the following functions:

1. Accepts approved documents for distribution and storage.
2. Verifies that the release and issue form (see Figure 10) is completely filled out and signed as required by the manual. A release and issue form should be prepared when a drawing, parts list, specification, etc., is submitted for issuance to production or other groups.
3. Retrieves filed documents to be revised or duplicated.
4. Makes sure that no document master is given to a requester without proper authorization, usually in written form.
5. Files all documents by number or type or both.
6. Posts all document activities on status records so that the location, change status, distribution, etc., of the data are permanently recorded.
7. Notifies staff member when his prints are ready for pickup.

The data control clerk is placed in a position of great responsibility because once a document is officially transferred to the facility's control, any damage, loss, or unauthorized revision will be the facility's or clerk's fault. Therefore, these clerks must have a strict and literal compulsion to follow procedures. This quality is necessary so that they cannot be persuaded to violate the system because a draftsman is in a hurry. A soft and compliant clerk can therefore undermine the integrity of the company's documents. If shortcuts are deemed necessary, then the procedure for taking them should be defined in the manual under "Special Situations."

REPRODUCTION OPERATOR

The reproduction machine operator is primarily responsible for duplicating and folding documents for distribution and filing. His secondary functions include equipment maintenance, record keeping, and care of sensitized materials and original documents. He also may stamp documents with usage data such as PRODUCTION, INSPECTION, VENDOR USE ONLY, ENGINEERING MODEL, and REFERENCE.

3.3 Records and Special Files: Keys To Control and Identification of Data

The records and special files of the data control facility play a key role in the control and status identification of documents released to it for protection and preservation. These include the following:

1. Release and issue forms
2. Status cards
3. Drawing number assignment log
4. Drawing log number book by project
5. Checkout cards
6. Cross reference index between documents and their change order documents

RELEASE AND ISSUE FORM: CRITICAL CONTROL DOCUMENT

The release and issue form shown in Figure 10 authorizes the data control clerk to issue a document for use by the staff or customer. It is usually prepared by the originator of the document and submitted with the document master or original.

Upon receipt of the form and original, the required copies of the document are made and distributed. The form and original are filed for future reference. Each time the document is revised, a new release and issue form may be required.

STATUS CARDS FOR LIFE CYCLE RECORDS OF DOCUMENTS

One status card is prepared for each document released. It is a record for the life of the document. The status card shows changes issued or in process and distribution data. It may also be used to give checkout data; e.g., the person revising the vellum. It identifies the document and gives the latest revision letter and date of issue. A sample status card is shown in Figure 11.

CASE IN POINT: Failure to keep status cards of drawings can be expensive as well as inconvenient. A small company didn't use status cards to identify the revision status of its

drawings. As a result, it was a common occurrence for the manufacturing and purchasing people to be ignorant of the latest drawing status when building or ordering parts because there was no record to indicate when a change to the drawing was in process. Therefore, fabrication and ordering of bad parts was a common occurrence.

DRAWING NUMBER ASSIGNMENT LOG PREVENTS MULTIPLE ASSIGNMENTS

When drawing number assignments are made by data control, the clerk must keep a record of the numbers or block of numbers issued so that they can be traced to the user and to prevent issuing the same numbers to more than one person. The drawing assignment log should contain the following information: drawing number block, person issued to, department, date of issue, and initials of clerk.

DRAWING NUMBER LOG BOOK: PERMANENT RECORD OF ASSIGNMENTS

The drafting supervisor must keep a log book of each assignment made. If the company operates on a project basis, a separate log book should be maintained. When the project is complete, the log book is submitted for filing to data control. Thus, a record will be available for reference if other drawing lists are lost. The log book should contain: project name, title of drawing, originator, date number was issued, and person issuing number.

CHECK OUT CARDS KEEP TRACK OF ORIGINALS

When a document is checked out of the facility, a check out card is completed with the name of the receiver, his department, purpose, authorization, and date of check out. This card is then placed in the document file for future reference. If a card file system is used for status records, a special color card should also be filled out and filed in front of the status card. Thus, anyone asking about the status of a drawing can be told immediately that it is being revised by Mr. Y of the D&D department.

CROSS REFERENCE INDEX BETWEEN DOCUMENTS AND CHANGE ORDERS

An engineer or draftsman may need to know what engineering change orders have been released against a particular drawing, or they may want to know what drawing is affected by engineering order 1050. A cross reference index can solve this problem. The first part of the index should list each EO in numerical order with the drawing number it applies to. The second part should contain each drawing number, its title, and the EOs released against it.

Since the data control facility is in the best position to maintain this file and to provide the information to requesters, it should be the source for this information.

SPECIFICATION AND STANDARDS FILE FOR IMPORTANT INFORMATION

Data control should maintain a current and complete file of industrial and military specifications and standards as required by the staff to do its work effectively. The documents required should be identified by title, number, date, and publication source on a document request form submitted to data control for purchasing. The inputs for these documents may come from D&D personnel, design engineers, standards engineers, production supervisors, and product assurance personnel.

Each document received by data control should be logged in on a specification and standards index list or rotary card file for easy referencing. The data recorded in the file should be in alphanumeric order by identifying number.

When documents are frequently required by the staff, two or three copies should be obtained. As with drawings and other data, sign out cards should be completed before releasing a specification/standard to a requester. This is important because someone may have an urgent need for the document and no other copy may be available. Therefore, data control must be able to retrieve the document when needed.

D&D personnel may find the use of a data control file troublesome and inefficient. If this is the case, a satellite file may be kept by the D&D department. However, the labor savings and convenience of having a separate file right in the department should be weighed against the cost of duplicated facilities and documents. Remember that floor space, cabinets, and filing and ordering expenses can add up to major added costs.

If the company requires thousands of specifications and standards, the use of microfiche or microfilm can reduce space requirements and operating costs significantly.

CASE IN POINT: The Wisconsin Department of Transportation eliminated a $20,000 purchase of file cabinets by implementing a micofilming program. The department reduced a 60-year accumulation of paperwork to the space of a regular file cabinet.

Flooded with old records, the Division of Highways started microfilming several years ago. Today, most operations involve use of miniaturized records. Another advantage of this system includes floor space savings. A 4 inch square, ½ inch thick roll of microfilm replaces one standard file cabinet.[9]

DISASTER FILES FOR PRESERVING ENGINEERING DATA

The ever-present possibility of fire, flood, hurricane, or explosive damage to the building requires a disaster file to protect the large investment in engineering, architectural, design, and drafting labor. But in addition to labor, a disaster file helps protect ideas and concepts that may also be destroyed.

Many companies microfilm drawings and other data as soon as they have been released by engineering. Microfilm rolls are used to store the data, with about 500 documents being stored per roll. Microfilm rolls are then delivered to a distant location for safekeeping. A file is maintained to identify the document microfilmed and its location in terms of the disaster facility and the actual roll number containing the document.

PURGING DATA TO CONSERVE SPACE AND COSTS

There is a time for most data to be purged from the company files. From an absolute viewpoint, a document should be eliminated from the company when it has no significant present or future value because storage space can get out of hand as documents and file cabinets accumulate over the years. While most managers agree that documents should be destroyed under this condition, determination of "no significant value" is not so easy. However, some other criteria are available to help in this decision such as general government and industry recommendations for retention periods of different kinds of data. Sample retention periods for some types of documents are given in Table VII as guidelines. However, note that State retention periods vary considerably; therefore, contact local departments for specific requirements.

Note that the data storage cycle requires three steps: 1. active retention file; 2. transfer to inactive file; and 3. destroy data. It is wise to let the staff know which types of documents are retained permanently, so that they can clean out data which is of no immediate value but may be needed for reference in the future.

Table VII
Document Retention Periods*

Type of Document	Retention Period	Remarks
Correspondence	5 years	
Directives from officers	Permanent	
Forms	Permanent	
Systems and procedures	Permanent	
Contracts	Permanent	Contracts Dept.
List of materials	2 years	
Drafting records	Permanent	
Drawings	2 years	
Inspection records	2 years	
Laboratory test reports	Permanent	
Memos	After completion of project	
Product, tooling, design, engineering research, and specification records	20 years	
Quality reports	Permanent	Quality assurance dept.
Work orders	3 years	
Work status reports	After completion of job	
Inventory reports	1 year	
Requests for services	1 year	
Requests for supplies	1 year	
Applications, changes in status, and terminations	50 years	Personnel Dept.
Attendance records	7 years	
Job descriptions	2 years	
Time cards	3 years	
Training manuals	Permanent	
Union agreements	3 years	
Negatives	5 years	
Photographs	1 year	
Reproduction records	1 year	
Price lists	When obsolete	
Purchase orders	3 years	Purchasing Dept.
Purchase requisitions	3 years	Purchasing Dept.
Quotations	1 year	
Presentations and proposals	Permanent	
Building drawings	Permanent	
Occupancy certificates	Permanent	

Property deeds	Permanent
Building maintenance and repair records	10 years

*For detailed retention information, see "Guide To Record Retention Requirements," published by the US Government Printing Office. This report can be obtained from the Public Documents Distribution Center, 5801 Tabor Ave., Philadelphia, Pa. 19120 for about $1.

The data given in this table is for illustrative purposes only; refer to local, state, and Federal codes for exact retention periods.

3.4 Auditing Operations for Maintaining Top Performance

Design and implementation of the data control operation is a primary task. However, if correct control and use of data is to be assured, a continuing program of surveillance of data control operations is needed. This inspection can be made by the data control supervisor or by an independent auditor such as quality assurance. A checklist should be prepared as a guide for specific areas to be investigated.

3.5 Types of Releases For Controlling Usage Of Data

A document release type refers to the approved application or usage of a document. Several types of document releases are used in industry. Production, limited, reference, and checkprint are a few of the common designations stamped on various documents to identify their usage level. These and other designations are described next.

PRODUCTION RELEASE

A "production" release refers to a document that can be used to build, modify, test, or purchase an item or product to the company's highest standards. It is only used when the design has been well-defined and the prototype has been built and tested. Note that for tighter control, the use of a pink (or other color) print paper may be used to positively identify a drawing that is acceptable for use by production personnel.

LIMITED RELEASE

A "limited" release applies to documents to be used for building experimental or preliminary hardware or for items to be used internally. An engineering model, tooling, or fixture fall into this category. The requirements, therefore, are not as stringent as for a production release. For example, quality, detail, and signature approvals are subject to less demanding standards.

REFERENCE RELEASE

"Reference" release indicates that the document is to be used for information only and that manufacturing, procurement, or testing operations cannot be performed using the document.

ADVANCE RELEASE

When working under tight schedules, pressure may be applied to production or purchasing to begin work on an item as soon as the basic configuration is defined. Thus, the project engineer may want a drawing to be released before it is complete. If this is feasible, the drawing may be released by data control with an "advance" release marking on the drawing. The user is therefore notified that the document is not complete although he can begin working on portions of the item that are completely described.

An "advance" release drawing or parts list should be replaced with a "production" document as soon as possible. Hopefully, in a few days or weeks. To avoid excessive delays, the data control clerk can flag the drawing status card with a red marker to indicate that the document is in an "advance" release state. Thus, a reminder can be sent to the project/product manager that this document has not yet been released for production.

VENDOR USE ONLY

To avoid inadvertently building a part that is slated for manufacture by an outside source, the marking "vendor use only" or "procurement" is used to indicate that a document is not for internal manufacture.

CHECKPRINT

A "checkprint" release refers to a document that has been completed by D&D and is ready for checking. A checkprint is never used for production, testing, inspection, or procurement.

The originator and D&D supervisor are responsible for indicating the correct type of release to data control. Data control cannot be responsible for this function and must receive written instructions defining usage data for each document released to it. Refer to Figure 10 for one method of identifying the release category. This release category is stamped in a clear area near the title block of the print.

3.6 Cost and Efficiency Factors

It should be clear to most people in industry that the control of documents is not a luxury but a necesstiy. With over 50 million drawings being prepared annually, the scope of the task for industry is a great one. Failure to effectively control drawings and other data increase labor costs or lose sales in the following areas:

1. A product is built to an incorrect print because the original was changed without telling production, procurement, or inspection about the change or issuing new prints.
2. Wasted time of draftsmen, engineers, and clerks increases when they spend hours trying to locate a drawing or parts list that has been misfiled or left in someone's desk.
3. Reproduction operators work longer hours making duplicate copies of documents which have already been printed but lost.
4. Draftsmen waste time redrawing vellums that have been torn or soiled because untrained personnel have been free to handle them.
5. A product is delivered late to a customer because the right drawing or test procedure couldn't be found until the responsible engineer returned from his vacation or business trip.
6. Sales are lost because a copy of a drawing showing a technical breakthrough is obtained by a competitor through sloppy filing and control of documents of engineers and draftsmen.
7. Originals have to be redrawn because fire or water has damaged them beyond use.

Minimizing the number and complexity of forms can reduce costs. By keeping them simple, high paid draftsmen, designers, and engineers do not have to spend as much time preparing and processing release and issue forms, reproduction requests, and drawing change notices. In addition, their labor costs can be reduced by having lower paid data control clerks fill out and process these forms as much as possible. This increases the data control facility operational costs but benefits the company by reducing total operating costs.

Using advanced or automated filing and retrieval systems can save the company money when the work-load or conditions justify their use. Automated systems can also reduce storage costs, waiting time of the requester and data control clerk, and reproduction operator labor.

The benefits of centralization were discussed in Chapter 2. They apply to the data control facility also. Therefore, whenever feasible, the facility should be located in one place closest to the greatest number of users. When a company is so large that this is not practical, satellite facilities can be established, but should be under direct control of a central unit. A positive and fast method of communicating status of data between the central and satellite units must be established to assure coordinated activities and properly updated records.

Although the staff members of the data control facility do not need extensive education and can learn the operations involved in a few months, they represent an important investment because one mistake can cost the company hundreds or thousands of dollars. Therefore, it is important to maintain a good esprit de corps and satisfied staff to minimize turn over with its additional training and error costs.

The use of dash numbers to identify non-interchangeable configurations can reduce record keeping expenses when compared to the use of mono-drawing number systems. For example, when a mono drawing identification system is used to indicate that an engineering change has made the item shown on the drawing non-interchangeable with its predecessors, the following work will result:

1. A new drawing status/history card must be prepared for the new drawing number created.
2. A new file folder must be prepared.
3. A new drawing number has to be issued and recorded in the log book.
4. Paper work must be issued to obsolete the old drawing number if this is required by engineering.

If the drawing number remains the same and only the dash number is changed, as described in Chapter 5, to indicate non-interchangeability, the above operations can be either entirely eliminated or drastically reduced. Since the above operations can easily consume a ½ hour per change, the savings can become significant when multiplied by several hundred or thousand changes per year.

Other cost saving techniques include reduction of space used for storing documents. For example, if the cost per square foot of space is 50¢ a month, a reduction of space requirements by 300 feet can save the company $1800 per year. Space reductions can be achieved by the following:

1. Use tall open shelf files, with retractable doors if needed.

CASE IN POINT: Universal Oil Products Transportation Equipment Division saved 1600 square feet by converting from 55 standard 5-drawer file cabinets to 9 shelf-type storage units. A major portion of the space savings was in the

print crib where this company files more than 45,000 prints covering the many aerotherm seat models and their component parts. The shelf files that replaced the regular cabinets hold about 5000 prints and use only 1/3 the floor space. Other advantages of this system are more readily accessible prints and easier identification of documents.[10]

2. Use lateral files that open sideways instead of the normal lengthwise fashion.
3. Use thinnest paper practical for prints to reduce filing space requirements.
4. Use reduced-size prints to conserve filing space.
5. Transfer files to a low cost storage area whenever they are reclassified from active to inactive category.
6. Periodically check files for excessive copies and destroy them.
7. Reduce storage, filing, and retrieval labor and time with new technology.

CASE IN POINT: GTE Sylvania reduced paperwork manpower by 77 percent by using microfilm for storage, retrieval, filing, and editing of engineering data. These results were achieved by installing a Recordak microfilmer, film processor, reader-printer, image control keyboard, and a reader.

Documents are microfilmed in random sequence as they come into the department and are placed into an automatic feeder of the microfilmer. Before filming, an index number is stamped on the document. Identification information is put onto EDP cards for retrieval of correct microfilm roll containing the desired document. This document is viewed on a screen and a copy made when needed.[11]

3.7 Computer-Aided Engineering Data Control Systems

For medium and large size companies, computer-aided engineering data control systems offer several possible advantages. These include:

- More efficient control and retrieval of large volumes of data.
- Real time response to inquiries on the status of data; e.g., in drafting, release, change pending, or change cancelled.
- Flexibility in reporting status of a particular job or program.
- Less clerical support and related forms and reports.

A possible configuration for a data control system could include a processor, several CRT display/input terminals, memory banks, a printer and a card reader. Input of data to the system could be controlled by one data control clerk using an input-output CRT terminal. The staff would use one of the several terminals distributed throughout the company to obtain the information it needs on a real-time basis. If special printed reports were needed of the desired data, these would also be provided.

The cost of such a system, of course, depends on its sophistication and scope. In one medium-size manufacturing company, a system was developed that cost about $60,000 a year. This included six video (CRT) terminals, a processor, a printer, a card reader, disk and tape memory, and recurring software changes. The savings in personnel, paper, and forms was determined to be $67,000 thus providing an annual net savings of almost $7,000 plus a faster, real-time data control system that met the needs of the engineers, administrators, and program managers.

4

4.0 How to Illustrate Product Requirements and Integrate Key Engineering Data

Drawings and associated engineering data depict the requirements of a product by pictorial and textual means. Drawings, of course, rely primarily on graphic or pictorial representations of the product's configuration and design. Other types of engineering data use textual material (narrative and tabulations) to define specific areas not easily described by a graphical approach.

The object of this chapter is to describe the various types of data in use and to identify their salient features. Keep in mind, however, that there are many variations in the formats, styles, and contents of these documents. Each company often has its unique data requirements. When work is done for the government, however, a company may be required by contract to follow the requirements specified in a standard or specification such as MIL-STD-100 or MIL-D-1000. Therefore, the use of company formats may not always be acceptable.

The most common use of drawings is to manufacture or build a product. However, drawings can have other uses such as design evaluation and interface control. Table VIII defines several categories.

Table VIII
Drawing Use Categories*

Category	Function or Use
Design evaluation	To evaluate a design or to document research and development work.

Interface control	To evaluate and control interfaces among interrelated components, subsystems, etc.
Interchangeability control	To identify and classify items to control their interchangeability, substitutability, and replaceability.
Logistics Support	To support spare parts ordering, cataloging, item identification, source coding, and standardization functions. This category of drawing is not used to purchase items.
Installation	To allow installation of parts, components, equipments, subsystems, and systems.
Procurement (Identical items)	To permit competitive purchase and manufacture of items that are essentially identical to original items (including identical repair parts).
Procurement (Interchangeable items)	To permit purchase of items that are interchangeable with (but not necessarily identical to) original items such as interchangeable assemblies or interchangeable repair parts.
Maintenance	To provide information to maintain and overhaul equipment, subsystems, or systems.
Customer/company manufacture	To provide information for the manufacture of an item by a customer or company.

*Adapted from MIL-D-1000, Drawings, Engineering and Associated Lists

BENEFITS AND APPLICATION OF THIS CHAPTER

This chapter presents a comprehensive survey of engineering data used in manufacturing, structural, and architectural fields. Its specific benefits include:

1. Provides broad review of drawing requirements, format, content, notes, and uses.
2. Covers a variety of document types: drawings, specifications, and lists commonly used in industry.
3. Provides sample drawings of many kinds of drawings.
4. Analyzes the various types of information required for drawings besides basic graphics data.
5. Presents specific drafting techniques for reducing the costs of preparing drawings.

The material presented in this chapter covers a wide range of projects and few companies are required to prepare all the document types described. Many companies may never prepare piping diagrams or printed wiring board drawings. Others may never prepare separate parts or data lists. However, familiarity with this information provides the engineer, designer, or draftsman of any company with a background that can (1) provide ideas on new document types that may be profitably applied to his organization; (2) provide understanding of drawing types used by vendors, subcontractors, or the government agency that the staff work with, (3) enable him to understand the drawing requirements of a company that he has just joined, or (4) allows a more efficient transition into new areas previously not a part of his company's domain.

4.1 How To Prepare Engineering Drawings

Drawings are made in various formats, sizes, grades, and types. While each company uses a different format, most drawings consist of the same basic elements: pre-printed borders, title block, and revision block. See Figure 12. When a larger size drawing is used, zone symbols enable the user to locate the revised portion of the drawing. Capital letters starting with "A" designate specific segments of the drawing in the vertical direction. (These letters start from the bottom of the sheet and go up.) Numbers, starting from right to left, identify areas of the drawing in the horizontal direction. Thus, by specifying a letter and number (for example, A4), the user can be directed to the approximate location of the revision or subject of interest.

The field of the drawing refers to the area within the borders and excludes the revision and title blocks. This is the area used to depict the product or item. Besides pictorial, dimensional, and tabulated data, a list of materials and parts may be added directly over the title block. Find numbers are also used to cross reference items or parts in the drawing to the parts list, which contains detailed information on all the parts. Notes defining special requirements such as "remove burrs and sharp edges," are also contained in the field of the drawing.

Note that large arrows are located at four points midway along the horizontal and vertical sides of the sheet. These help the operator center the sheet when microfilming is needed.

Although the title block format is rarely identical from company to company, within the company it should be the same to assure uniformity of appearance and content. The sample title block shown in Figure 12 consists of four parts:

- Usage block
- Identification block
- Approval block
- Standard note block

USAGE BLOCK: The usage block or application block gives the usages of the item on the drawing. The entry on the left is the drawing number of the

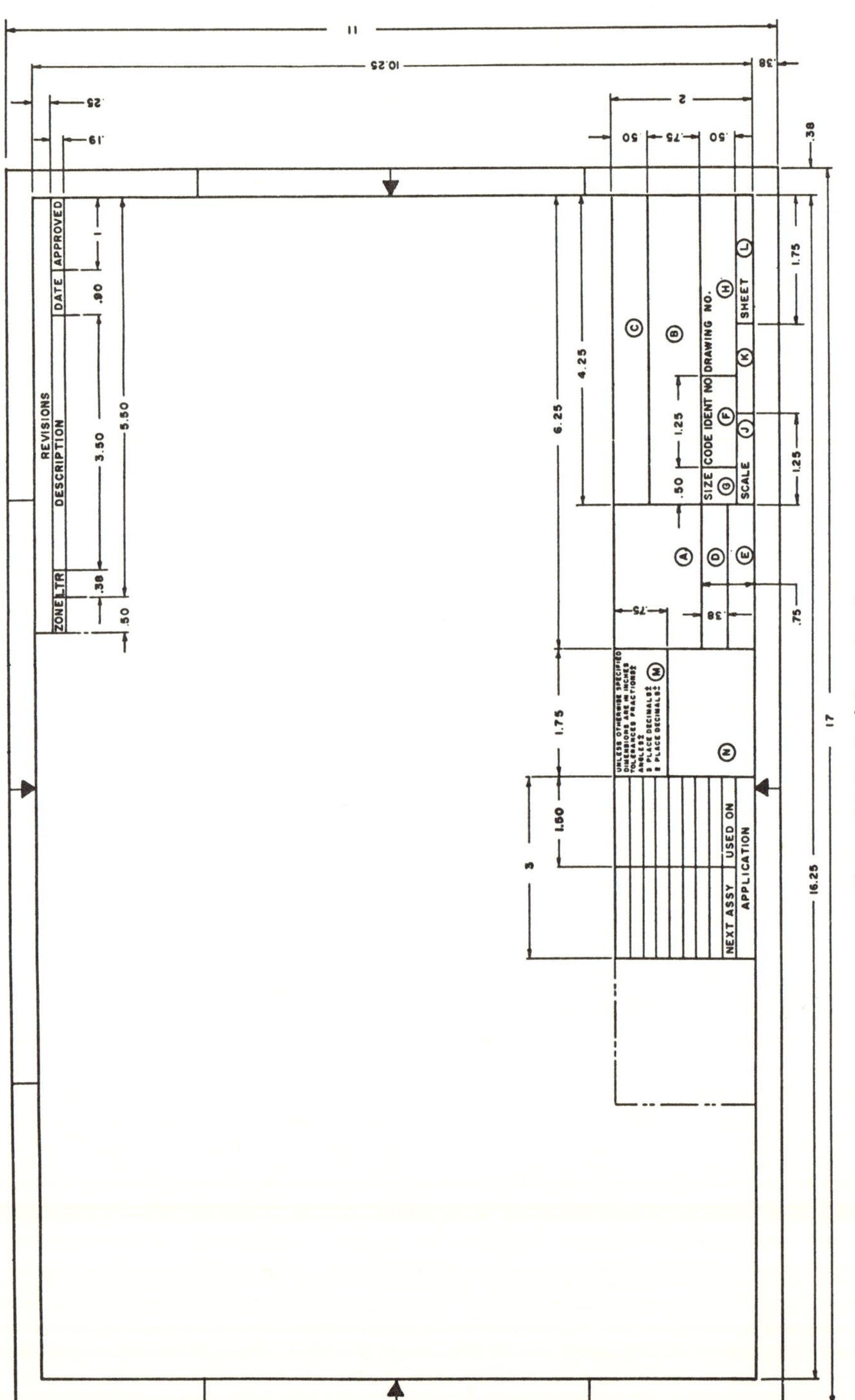

Figure 12 Typical Drawing Format

assembly into which the item goes. The next column identifies the end item or model number of the product into which the item goes. This block contains "FINAL" in the final assembly drawing. Note that several entries may be made in these columns when the item is used in different end items or assemblies.

IDENTIFICATION BLOCK: The identification block gives the name of the company and its location, the title of the drawing, its number, size, scale, and sheet number. The code identification number is used when government design and production work is done. This 5-digit number is assigned by the government to uniquely identify each company and its major divisions. A contract or job number entry block is also used.

DRAWING APPROVAL BLOCK: Drawing approval signatures are recorded in the approval block. The number of signatures is determined by the type of drawing and company policy. Some drawings may only need the signatures of the draftsman and project engineer. Others may require signatures of the draftsman, checker, designer, project engineer, manufacturing engineer, and quality assurance representative.

STANDARD NOTE BLOCK: A standard note block is also often used to indicate certain commonly required information so that the draftsman doesn't have to waste time repeating it on each drawing he makes. Typical information includes: "Unless otherwise specified, dimensions are in inches" and "tolerances for three place decimals are ± .005 inch."

REVISION BLOCK

The revision block contains a summary of the changes incorporated into the drawing. It includes zone identification, change letter, description, date, and approval signature columns located at the upper-right-hand corner of the drawing. It may also contain an effectivity column for recording the end item serial numbers affected by the change. As was mentioned, the change letter column identifies each change by a capital letter in sequence of incorporation. Thus, "A" indicates the first revision made, "B" the second, "C" the third, and so on. The description column contains a brief summary of the change made. For example,

"Bend radius 1/16 R was 3/32 R" or
"Sheet number was changed from 3 to 4."

The date of the revision is recorded in the column shown in Figure 12. To make the change official, the signatures of the persons originally approving the drawing are repeated in the approved column. The number of signatures depends on company policy but should be at level adequate to determine that the change has been made correctly and that its impact on the end-item has been properly evaluated.

Sufficient space should be allowed for adding revision notes as the drawing is changed. Thus, the drawing should be laid out so that it does not occupy the area immediately below the revision block. Guidelines for how much space to leave follow:

Drawing Size	Inches Below Revision Block
A	1.5
B	3
C	3
D	5.5
E and larger	7

When required, space should also be provided on drawing for reference drawings, customer's name, information required by building codes and regulatory agencies, and special data.

DRAWING SIZES

Drawing sizes vary considerably, depending on the scope and complexity of the item to be depicted. (Standard letter designators are used to indicate the dimensions of drawings.) Sometimes a standard sheet size paper is sufficient. Other times roll sizes a few feet wide and several feet long are needed. While the smallest size drawing is preferable because of ease of handling and the lower cost of reproduction, use, and storage, drawing sizes should be adequate for convenient and clear presentation of the item. In addition, standard sizes should be used to simplify ordering, folding, and storage operations. Some standard drawing sizes are listed below:

Drawing Size Letter Designator	Dimensions in Inches
A	8½ by 11
B	11 by 17
C	17 by 22
D	22 by 34
E	34 by 44
F	28 by 40
G	11 by 42 to 144 (roll)
H	28 by 48 to 144 (roll)
J	34 by 48 to 144 (roll)
K	40 by 48 to 144 (roll)

Note that because the above drawing sizes are available, they need not all be used. Each company has to decide which sizes are required that are most economical. Some companies limit drawing sizes to A through E and use multiple sheets if the entire drawing can not be fit on the sheet. This method, of course, may not always be practical, and much larger sheets may be needed.

MULTISHEET AND BOOK FORMATS

Many drawings consist of only one sheet. However, others have two or more sheets, especially if they are in book format. A book format is simply a group of standard 8½ by 11 inch sheets identified by a common drawing number and stapled together at one corner. A cover sheet, separate note sheet, and a few drawing sheets usually constitute a book form drawing. This is a practical document to work with when items produced by a company can be depicted on an A or B size sheet. The book format offers convenience of handling, storage, and reproduction.

Note that regardless of whether standard multisheet or book format drawings are used, each sheet must be identified with the drawing number, revision letter, and its individual sheet number.

DRAWING TITLE

A drawing title is the name by which the part is known. A drawing title consists of the item's name and additional modifiers required to identify it more precisely. Its basic format is as follows:

Note that the title is always in upper case letters and usually consists of 3/16 inch high letters.

A drawing title should be as simple and brief as feasible. Its brevity, however, must be determined by its ability to describe the item and distinguish it from other parts in the product or system. Although abbreviations should be avoided, they can be used to reduce the title length, especially in its modifiers. The prime item name should not be abbreviated except for ASSEMBLY (ASSY), SUBASSEMBLY (SUBASSY), or INSTALLATION (INSTL). Don't use periods after abbreviations unless their omission will cause confusion.

The prime item name should be selected to establish the basic concept of the item. It describes the part and usage of the part, but does not describe or identify the material used or method of fabrication. Also don't use trade names, trade marks, or copyrighted names and processes in title.

The second part of the title amplifies the first part by telling what the part is (shape, structure, etc.), its function, or its location. This part of the title is separated from the first part with a dash having spaces on both its sides. Single word modifiers or compound words may be used for providing this information. Modifiers, whether in the first or second titles, are separated with commas.

DRAWING BORDER DIMENSIONS

The exact border dimensions used are not sacred. However, consistent borders should be used for all company drawings as defined in the D&D manual or some other standard. Some examples of drawing borders for various size drawing formats follow:

Drawing Size	Vertical Borders (in.)	Horizontal Borders (in.)
A	.25	.38
B	.38	.38
C	.50	.50
D	.50	.50
E	.50	.50

FIND NUMBERS (Also called Item Numbers)

When a parts list is used, each part or item shown on a drawing is identified with a find number for cross referencing it to the parts list. (If the part number is given next to the part, a find number is not used.) The number is necessary regardless of whether a separate or integral parts list is used. The find number (often referred to as FN or F/N) is placed in a circle or elongated circle on the field of the drawing and a line is drawn to the part it identifies. The same number is listed in the parts list, which contains a full description of the part; e.g., its part number, manufacturer, specification, and quantity required.

DRAWING NOTES

Supplemental information and instructions are given by notes in the field of the drawing. Notes can be placed close to the item of application or grouped separately on one side of the drawing and numbered consecutively. An example of notes follows:

1. This tolerance applies to all points on the surface of the assembly.
2. Steel stamp in ¼ inch characters: 123456-11.
3. Test to TP 123456.

Although notes may be placed anywhere on the field of the drawing that does not interfere with the item shown, a standard or preferred placement of notes should be identified for uniformity. Some general guidelines for listing notes follow:

* Record "NOTES: UNLESS OTHERWISE SPECIFIED" in the lower left hand side of drawing.
* Notes are numerically sequenced reading up from heading: "NOTES: UNLESS. . . ."
* List notes in their general order of application.
* List notes on the first sheet of a multisheet drawing.

When a note applies to a specific part of the drawing, a triangle or square is placed next to the applicable item with a reference number. This triangle and the reference is then repeated in the note area and the note is given next to it as illustrated before.

General notes (GN) apply to all affected items or operations shown on drawing and are listed above "NOTES: UNLESS. . . ." No triangles are used to enclose a general note number.

CIRCUIT REFERENCE DESIGNATORS

All components on a drawing should be identified by common/standard reference designators. These designators are placed as close to their components as possible and the component value is placed directly below it when possible. Samples of reference designators include:

R1—first resistor in series
C5—fifth capacitor in series
K1—first relay in series
Q4—fourth transistor in series
U10—tenth integrated circuit in series

When a part consists of two sections, each section is identified as follows: Q10A and Q10B (two independent sections of a transistor). Reference designators are usually numbered in order from left to right and from top to bottom of the schematic.

SCALES

Scales are used to give the user a basis for determining the true size of the item shown on a drawing. The scale is indicated by a fraction and placed in the title block as shown in Figure 12. The fractional method for indicating scale tells the user what the ratio of the size of the item shown is to the real size. For example:

1/1 indicates a full size drawing; i.e., the drawing is the actual size of the fabricated item. Note that some companies print "FULL" in the scale box instead of 1/1.
10/1 indicates that the item shown is 10 times larger than its true size.
1/4 indicates that the item is a quarter of the size of the finished item.

Construction drawings are usually drawn to a scale of a fraction of an inch per foot or inches per foot, hundred feet, rods, and miles. If a drawing has no scale such as a schematic, the word "NONE" is entered in the scale block. When drawings made to scale have certain dimensions out of scale, these dimensions are underlined in a wavy line as follows: 50

If a drawing has two or more scales, the scale of the main view is called out in the scale block "AND NOTED" added to cover other scales.

LETTER SIZE AND STYLE

Letter sizes for identification, administrative, and other information on drawings vary depending on the purpose of the information. A sample of one company's letter sizes follows:

Letter Type	Size (inch) Fractional	Decimal
Drawing number	1/4	.25
Title	3/16	.19
Subtitle	5/32	.16
Fractions and tolerances, dimensions, and notes	5/32	.16 (.18 on D, E, and J size drawings)
Designation of Sections & Detail Views	5/32	.16
Designation of Section and view letters A-A, B-B, etc. on all drawing sizes	1/2	.50

Lettering style should be simple, clean, and in capitals. For example, single-stroke Gothic letters without serifs are preferred by many companies. Lower case letters are used when special requirements apply. Straight or slant letters are usually permitted but only one kind should be used on a drawing. Spacing between lines should be greater than .06 inch.

PARTS LIST/BILL OF MATERIALS

A bill of materials is often an integral part of a drawing. As mentioned, it may appear over the title block and contains detail information on the items specified on the drawing. This information includes a description of the item, part number, vendor or manufacturer, applicable specification, and quantities needed. Sometimes the parts list is prepared as a separate document. Separate parts list preparation is discussed in the next section on associated engineering data.

The term "bill of material" is often used in the structural or architectural fields. This type of list includes stock sizes, kinds of wood, board measure, bolts, and other metal parts.

TYPES OF DRAWINGS

Many types of drawings are used in industry. They include detail, assembly, control, installation, and diagram categories. Other drawing types include chemical flow sheets, structural, architecutral, piping, and topographical. Each drawing type has a special purpose and thus serves to provide a complete package that describes the design, construction, and configuration of a product. The functions, requirements, and special features of these drawings are discussed next.

DETAIL DRAWING—A detail drawing portrays the complete requirements to build a single item. Two or three views of the item are shown so that the user can see the top and side configurations. The drawing gives a complete and exact description of the part's shape, dimensions, and construction. Accuracy limits, hole diameters, finish, material, and notes are given to cover all aspects of the item. Supplemental reference documents are also recorded on the drawing. A list of fabrication or machining operations may also be given.

Detail drawings fall into additional subcategories as follows:

- Monodetail: shows only a single part
- Multidetail: shows more than one part
- Tabulated: contains table of variable dimensions or other characteristics associated with the basic part configuration depicted on the drawing. Tabulated drawings are used when one drawing can be used for several different configurations of a part by simply tabulating certain dimensions or other characteristics. Refer to Figure 13.*

*Drawing samples given in following sections are from MIL-STD-100, Engineering Drawing Practices, 1965.

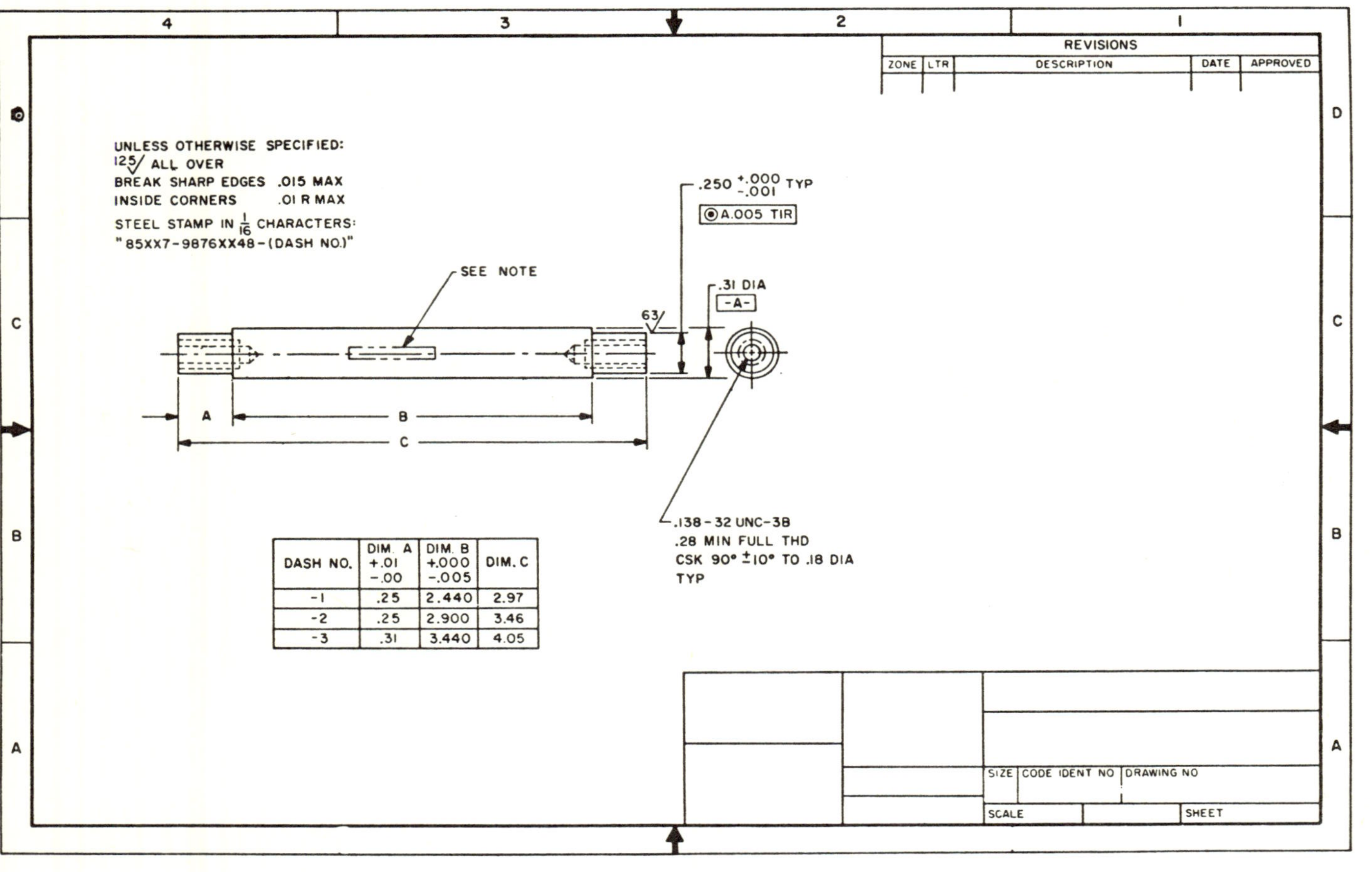

Figure 13 Tabulated Drawing

BENEFITS OF DETAIL DRAWINGS—Compared to assembly or multi-detail drawings, detail drawings offer the following benefits: (1) cleaner, less cluttered and easier to read and work from; (2) allow depicting a part that could not be readily presented in higher assembly drawing; (3) permit more accurate and complete description of item as compared to assembly and multi-detail drawings; (4) can be put on a smaller drawing sheet; (5) easier and faster to prepare and revise; (6) easier to build from and to inspect; (7) allow different parts to be made by different people, e.g., machinist, welder, or patternmaker; and (8) simplest form of working drawing. Detail drawings should be prepared whenever a working drawing of the item is needed and is not depicted elsewhere.

ASSEMBLY DRAWING—An assembly drawing (see Figure 14) shows the assembled configuration of two or more parts in their proper relative positions. It may also portray a combination of assemblies needed to make a higher level assembly. This drawing should contain references to separate parts and data lists, installation drawings, and wiring diagrams. Other information includes views for easily assembling all parts, dimensions, hardware needed to join parts/subassemblies, notes defining special requirements, specifications, and procedures. Assembly drawing subcategories are described next:

Top Assembly: shows highest assembly level of an item or product.

Arrangement assembly: identifies principal subassemblies and shows their interrelationships.

Detail assembly: shows details of one or more parts in addition to an assembly.

Subassembly: shows two or more parts attached to each other before their incorporation as a unit into an assembly.

Tabulated assembly: similar to tabulated detail drawing, except for containing assemblies instead of parts.

Matched parts: shows parts that are mated and must be used together and which must be replaced as a matched set. (See Figure 15.)

Photographic assembly: shows item by using half-tone photographs of the assembly with call outs for parts/subassembly identification.

Exploded assembly: shows elementary parts of assembly projected in space from each other.

BENEFITS OF ASSEMBLY DRAWINGS: Compared to detail and multi-detail drawings, assembly drawings have considerable value in saving time and labor. The following benefits can be derived from assembly drawings: (1) assembly drawings can make it easier for a technician to put the item together more accurately and faster than if he only had detail drawings available to work

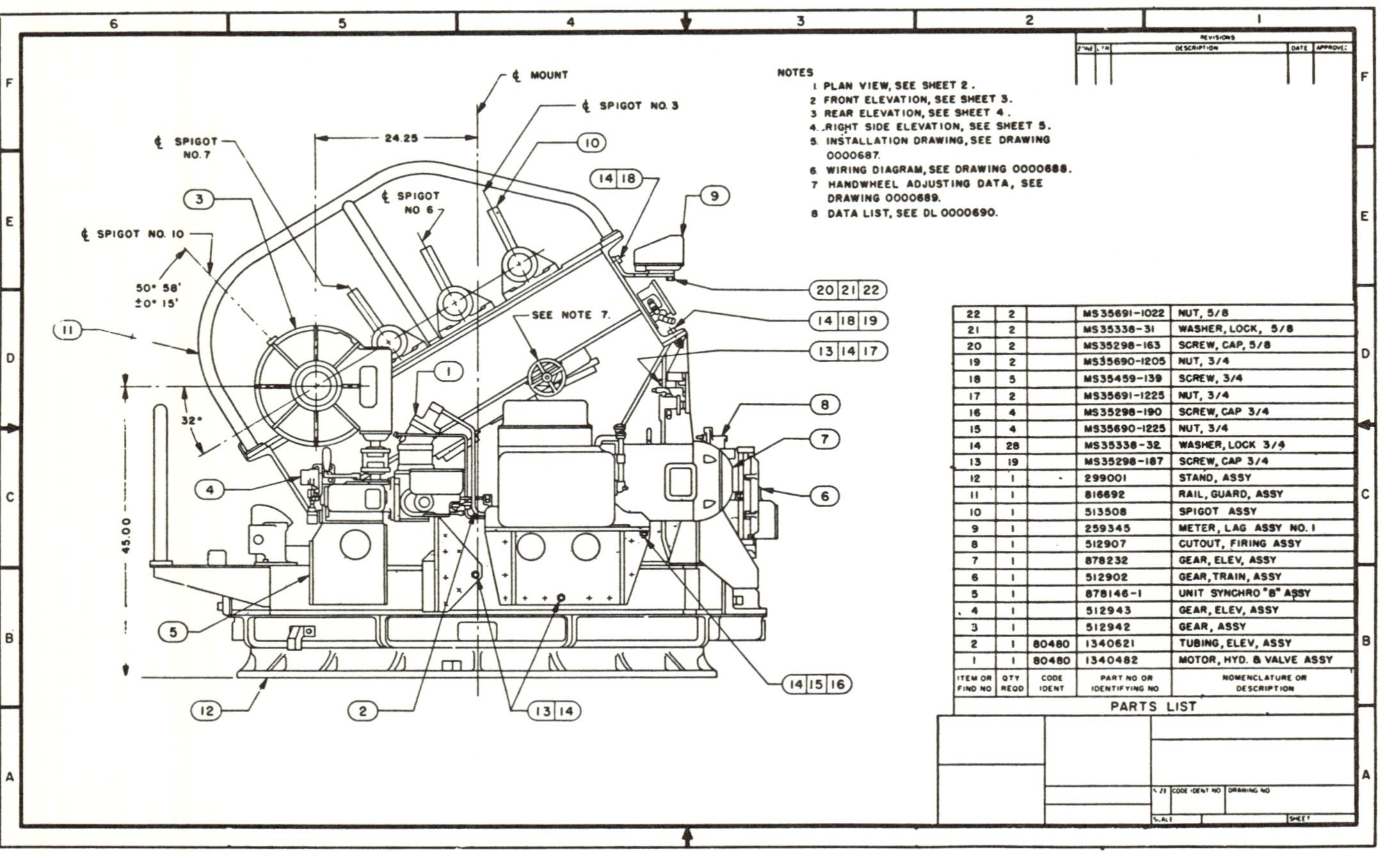

Figure 14 Assembly Drawing

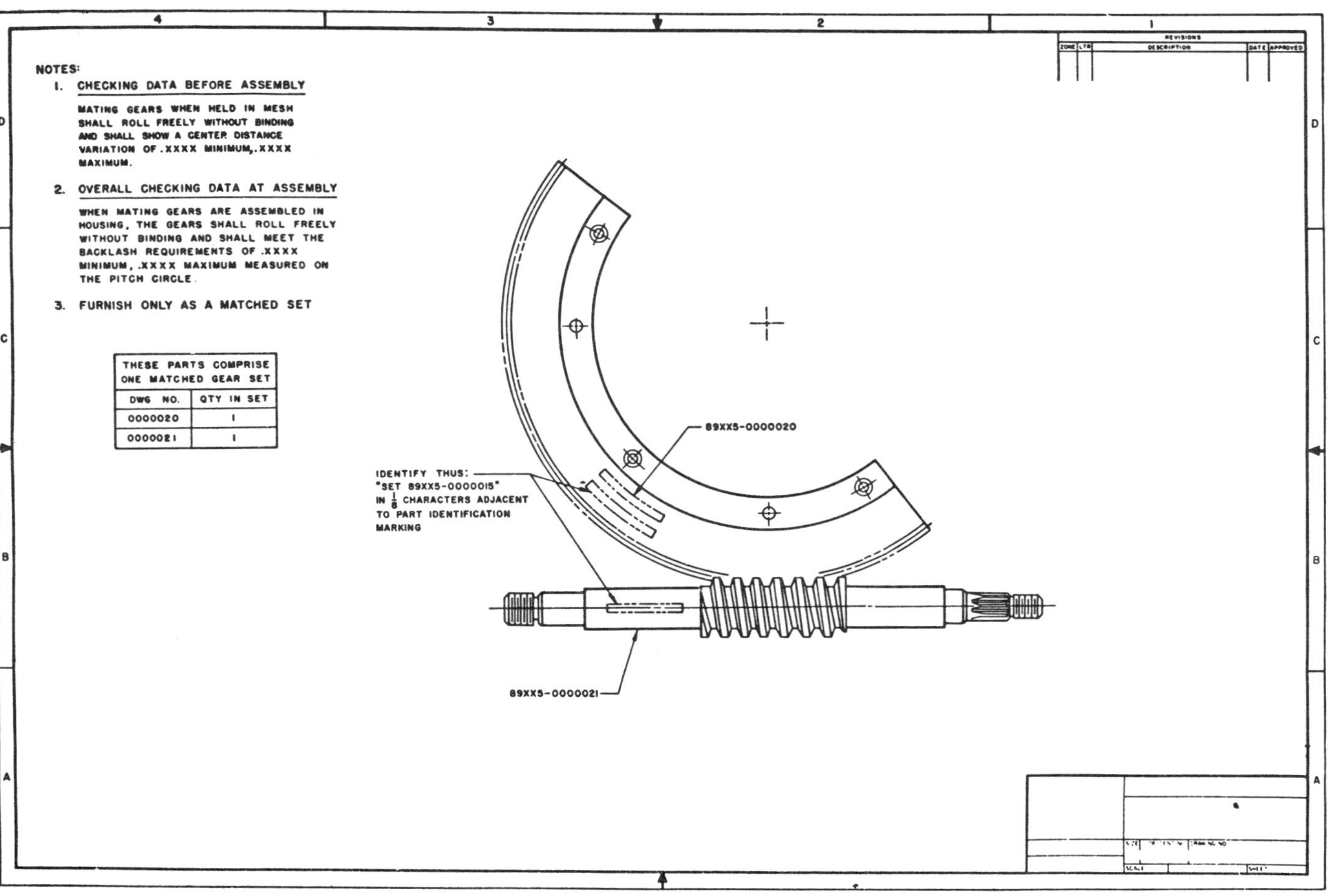

Figure 15 Matched Parts Drawing

from; (2) an assembly drawing can present a clearer picture of a device because information needed for detail drawings can be omitted such as hidden lines, extensive dimensioning, and many notes; (3) the preparation of an assembly drawing allows an excellent method of verifying the accuracy of detail drawings and can reveal clearance or fit problems; and (4) in many cases, it is cheaper to prepare one drawing of all the parts than to prepare separate detail drawings of each one.

Assembly drawings can be made for different purposes. Therefore in some cases, only one view of the assembled item is shown. In other cases, two or more views may be given.

INSTALLATION DRAWING—An installation drawing shows how and where an item is attached to a major piece of equipment or to its supporting structure. It also shows how the item is mounted in relation to other components, subassemblies, or assemblies. Data recorded on this drawing include dimensions, weight, orientation, mounting and location of mating connections, electrical connectors, cabling, points of adjustment, and reference documents. Other information may include center of gravity, location of nameplates, and safety notices. The amount of clearance needed to allow doors to be opened or modules removed may also be given.

BENEFITS OF INSTALLATION DRAWINGS—An installation drawing has several benefits. It helps people in attaching or connecting an item to another device faster, easier, and with fewer errors than from verbal or written descriptions. It also helps prevent injuries and can avoid damage to equipment.

This kind of drawing should be used when the installation is too complex to visualize without the drawing or when routing or physical placement of cables and parts cannot be easily or precisely identified without a picture.

CONTROL DRAWING—A control drawing is designed to control form, fit, and function requirements related to an individual part or a group of items. Its purpose is to assure that the parts will present no installation problems when combined with other items and that they will work properly as a part of the overall system.

Some of the areas covered in control drawings are:

- Configuration: size, shape, dimensions, finish, etc.
- Weight, power, center of gravity, and space restrictions.
- Cable connections.
- Access clearances.
- Performance requirements.
- Test requirements.
- Mounting techniques.

Several types of control drawings exist such as installation control, envelope, specification control, source control, and interface control. These drawings are described next:

Installation control: shows requirements for the installation and operation of a unit when combined with other parts or assemblies. Data presented include weight, mounting elements, size and area, access clearance, electrical and mechanical connections, and mechanical linkages.

Envelope: shows configuration, performance, and constraint requirements needed to perform detail design of the item. It also is used to give a potential customer a dimensional outline drawing of the product. Depending on its use, an envelope drawing may contain information on interchangeability, test, reliability, key or maximum dimensions, and installation requirements. Connectors and mounting techniques are also shown.

Specification control: shows the configuration and requirements of an item that is commercially available. (Refer to Figure 16.) The purpose of this document is to assure that the vendor supplies parts that meet certain minimum requirements, but it does not restrict choice of a vendor or detail design characteristics not covered by the specification. This control drawing prevents the vendor/manufacturer from making changes that could affect the acceptability of the part without changing the part number. The drawing shall include configuration, test, environmental, reliability, installation dimensions, and inspection data to assure that the desired item is supplied.

Specification control drawings are usually identified as such with .25 inch letters above the title block or as near to it as feasible. Suggested vendors for supplying the item are also listed with their associated part numbers.

Source control: shows existing commercial item that uniquely provides the functional or other characteristics needed for a critical application. It is similar to a specification control drawing in terms of requirements information. However, instead of a suggested list of suppliers, only suppliers with proven capability to supply the item are listed. Note that a source control drawing number is also considered a part number. This drawing is identified as "SOURCE CONTROL DRAWING" near the title block.

Interface control: identifies the physical and functional aspects of an item that affect the operation or installation of the item with its associated system elements. Specifically this drawing does the following:

- Establishes compatibility among associated units.
- Controls interface designs to prevent changes that make associated items incompatible.
- Transfers design information and changes to affected organizations.

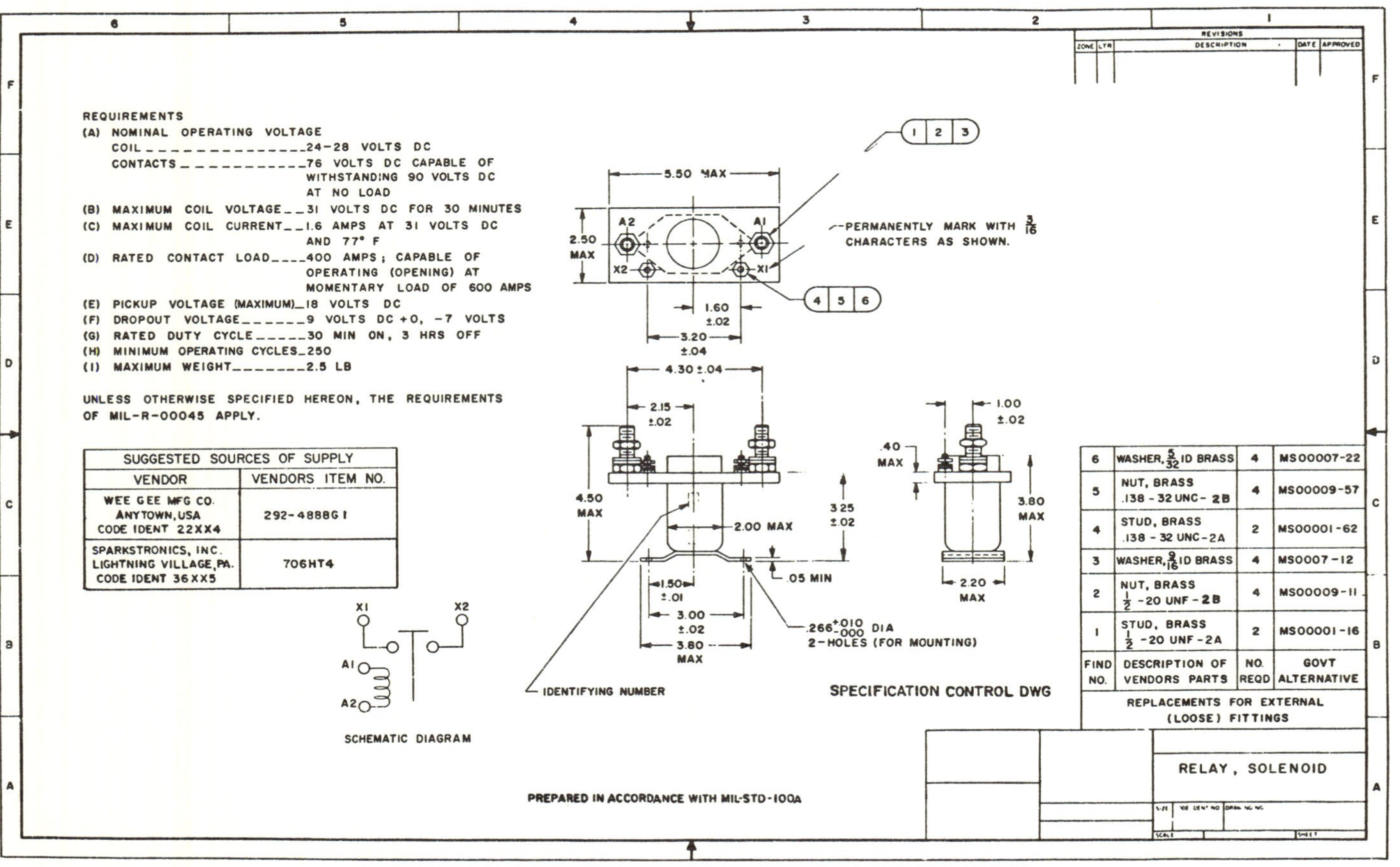

Figure 16 Specification Control Drawing

These drawings are identified by "INTERFACE CONTROL DRAWING" near the title block.

Correlation: similar to an interface drawing but restricted to physical and functional requirements among components within a subsystem. Therefore, interface requirements among associated subsystems are not used. "CORRELATION DRAWING" is added near the title block.

Altered item: shows changes to a standard part. Includes original vendor and part number in field of drawing. All details of the alteration must be given. This drawing is identified with "ALTERED ITEM DRAWING" near title block. See Figure 17.

Selected item: shows an existing item with additional data on selecting or restricting the item in terms of fit, performance, reliability, or tolerances. "SELECTED ITEM DRAWING" is recorded near title block. See Figure 18.

BENEFITS OF CONTROL DRAWINGS—Control drawings have several benefits compared to other types of documents: (1) help prevent future orders of a part from resulting in unusable parts because of vendor-made changes in form, fit, or function characteristics; and (2) help ensure that parts made by different groups or companies will mate or join properly. Another benefit is to assure that a device is installed properly in its next higher using assembly or facility.

Control drawings should be used when an item's configuration is critical to a project and the vendor's configuration control methods indicate a high risk of receiving an item that differs from your needs. Another condition is that the item will be reordered as a replacement part in the future. In this case the control drawing will ensure that it meets minimum requirements.

DIAGRAM—Diagrammatic drawings use symbols and lines to show characteristics, connections, and relationships of items comprising a component, assembly, or system. Scales and part thicknesses are not usually given. Typical diagrams include schematics, wiring, interconnection, piping, and logic diagrams. These are covered in the following paragraphs:

Schematic: shows the electrical or mechanical connections and functions of a particular circuit or mechanical device using a variety of symbols understood by a wide range of people because of mutual and official agreement as to their meanings. See Figure 19.

Wiring or Connection: shows electrical connections within an assembly or other component. The physical layout of the parts or components is usually included. Other items include color or alphanumeric coding of each wire, terminal points on each wire, and the length of wires. Input

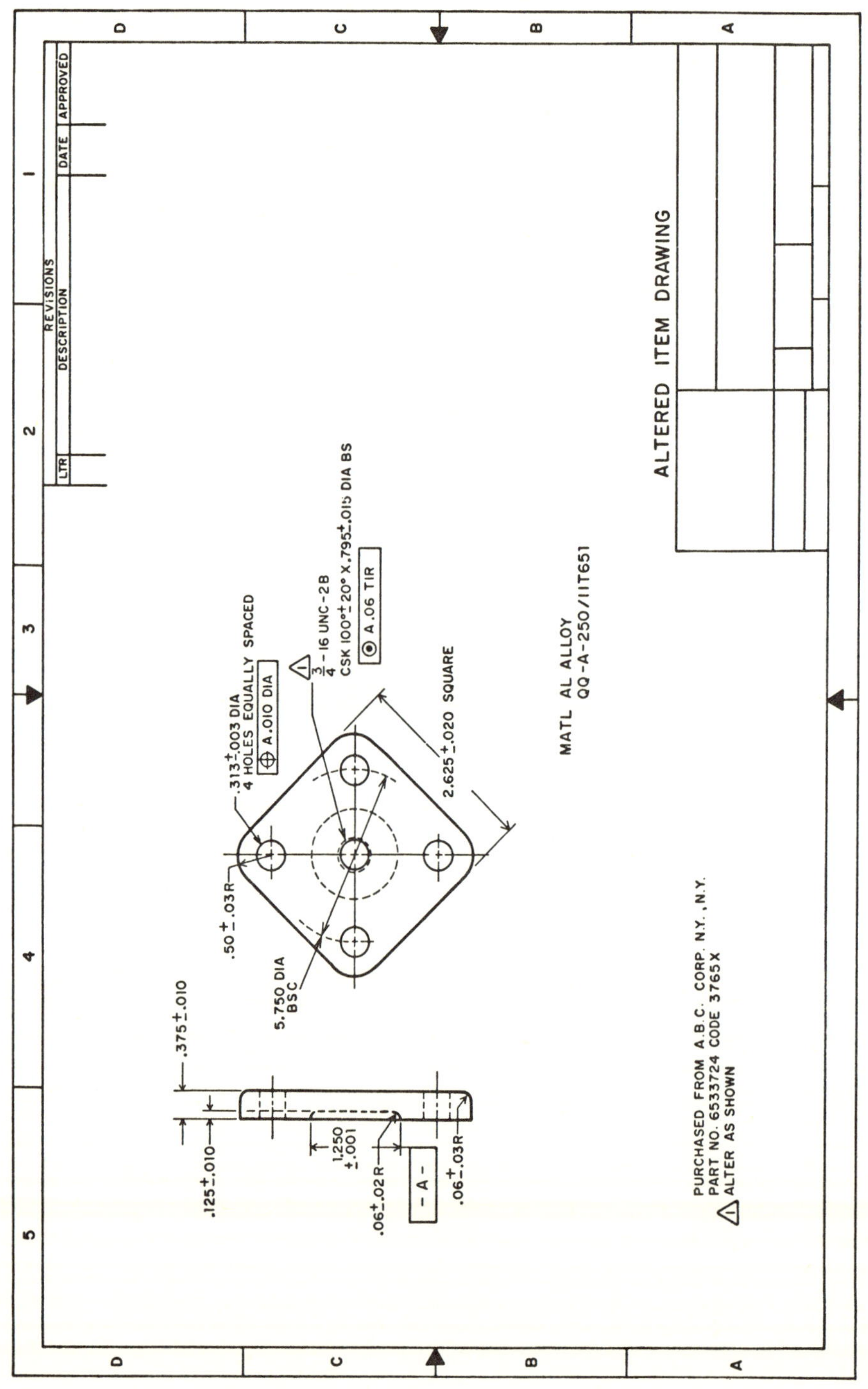

Figure 17 Altered Item Drawing

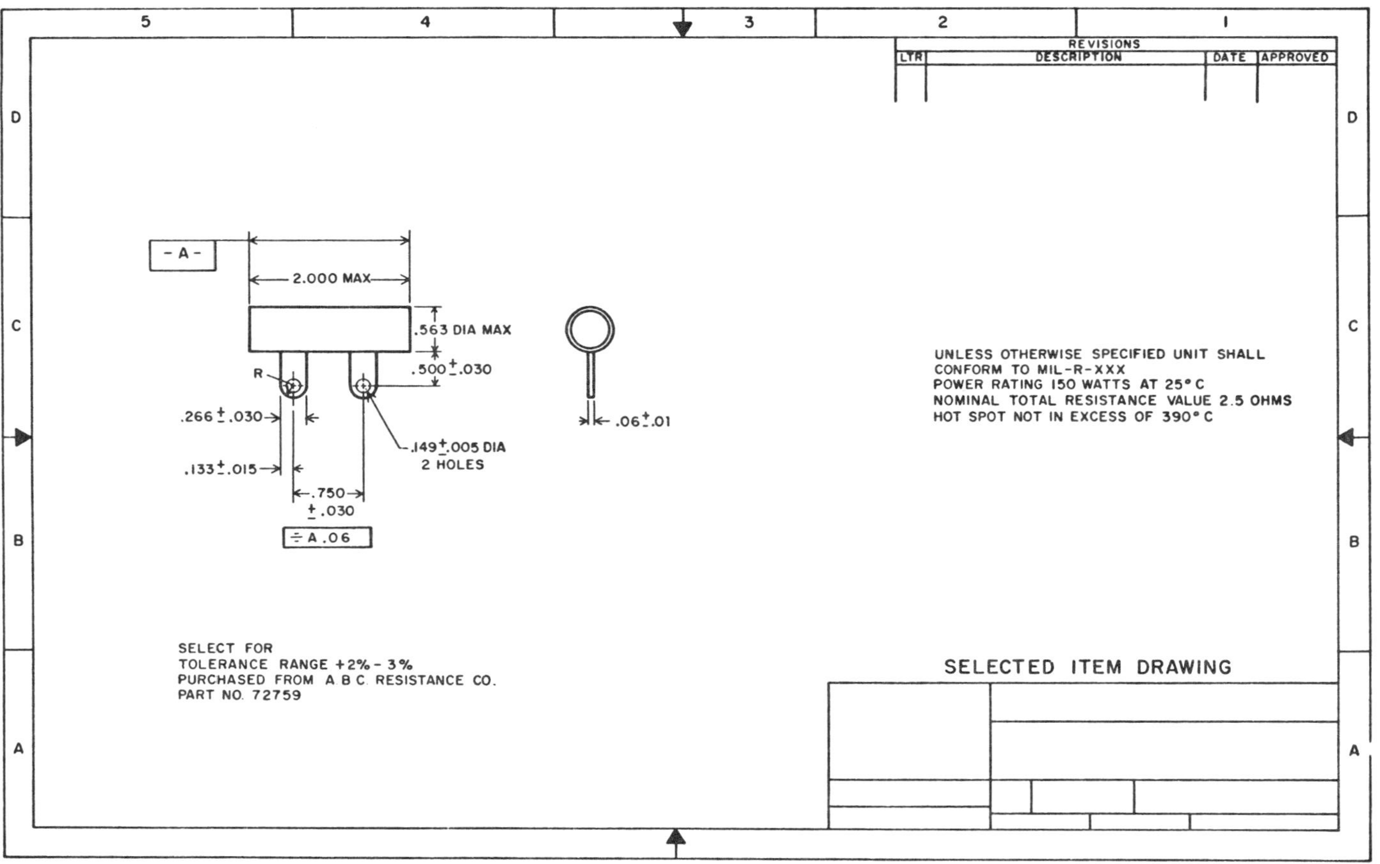

Figure 18 Selected Item Drawing

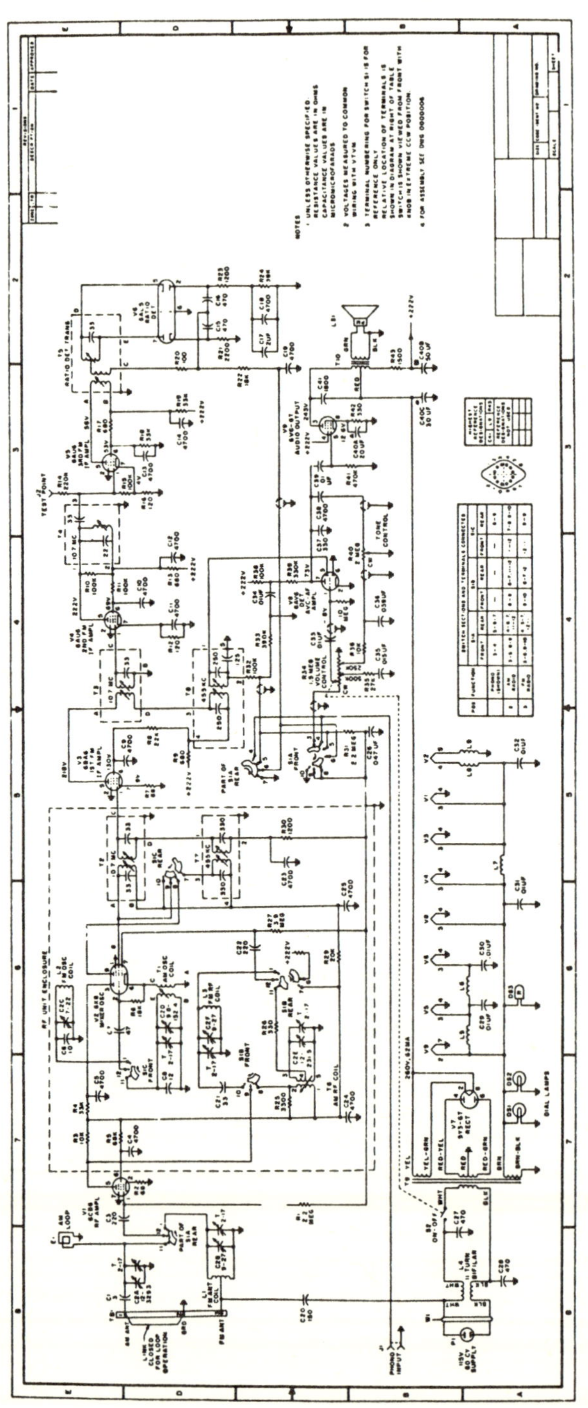

Figure 19 Schematic Diagram

power is usually shown coming from left hand side. Wiring diagrams are not needed if complete wiring information is given on the assembly drawing.

Interconnection: shows external connections among various levels of subsystems and systems.

Piping: depict interconnections of components by piping, tubing, or hose and sequential flow of fluids or gases in a system. These diagrams include pipes, fittings, valves, caps, couplings, reducers, unions, and flanges. Either single line or double line drawings may be made. However, the single line drawings are easier to prepare. Piping diagrams are used for power plants, pumping plants, refineries, heating systems, cooling systems, and plumbing systems. (See Figure 20.)

Block diagram or single line: shows by use of signal lines and symbols the path of an electric circuit. (See Figure 21.)

Logic: shows sequence and function of logic circuits by use of graphic symbols.

BENEFITS OF DIAGRAMS: The value of diagrams is in that they give a short-hand method of depicting how something works or is put together without regard to its physical characteristics. They also facilitate understanding that would not be possible by showing the exterior or physical configuration of the object. Other benefits are (1) improves the efficiency of the builder, operator, or maintenance man and (2) in many cases it is the only way to provide a means for understanding the construction or operation of an item.

CONSTRUCTION DRAWING—Construction drawings depict the design of buildings and structures. They include equipment, utilities, and other elements related to architectural and civil engineering designs. Construction drawings include erection, plan (see Figure 22), plot, and vicinity drawings. An erection drawing shows the procedures and sequence for erecting a facility or structure. A plan drawing provides a horizontal view of a building and shows the floor, foundation, and roof. A plot drawing shows a facility in relation to boundary lines, streets, walks, fences, and utilities. A vicinity drawing shows the relationship of a site to the surrounding area.

STRUCTURAL DRAWING—Structural drawings (see Figure 23) are prepared for structures such as buildings, bridges, and dams. Structural drawings depict the framework and supporting members of structures such as columns, floor slabs, roof trusses, and bridge trusses. Steel structural members and materials come in a variety of shapes, sizes, and weights. For example, bar, angle, plate, channel, tee, zee, and standard beam are typical shapes used in steel construction. Reference books are available defining the properties of the various materials, shapes, and combinations of shapes.

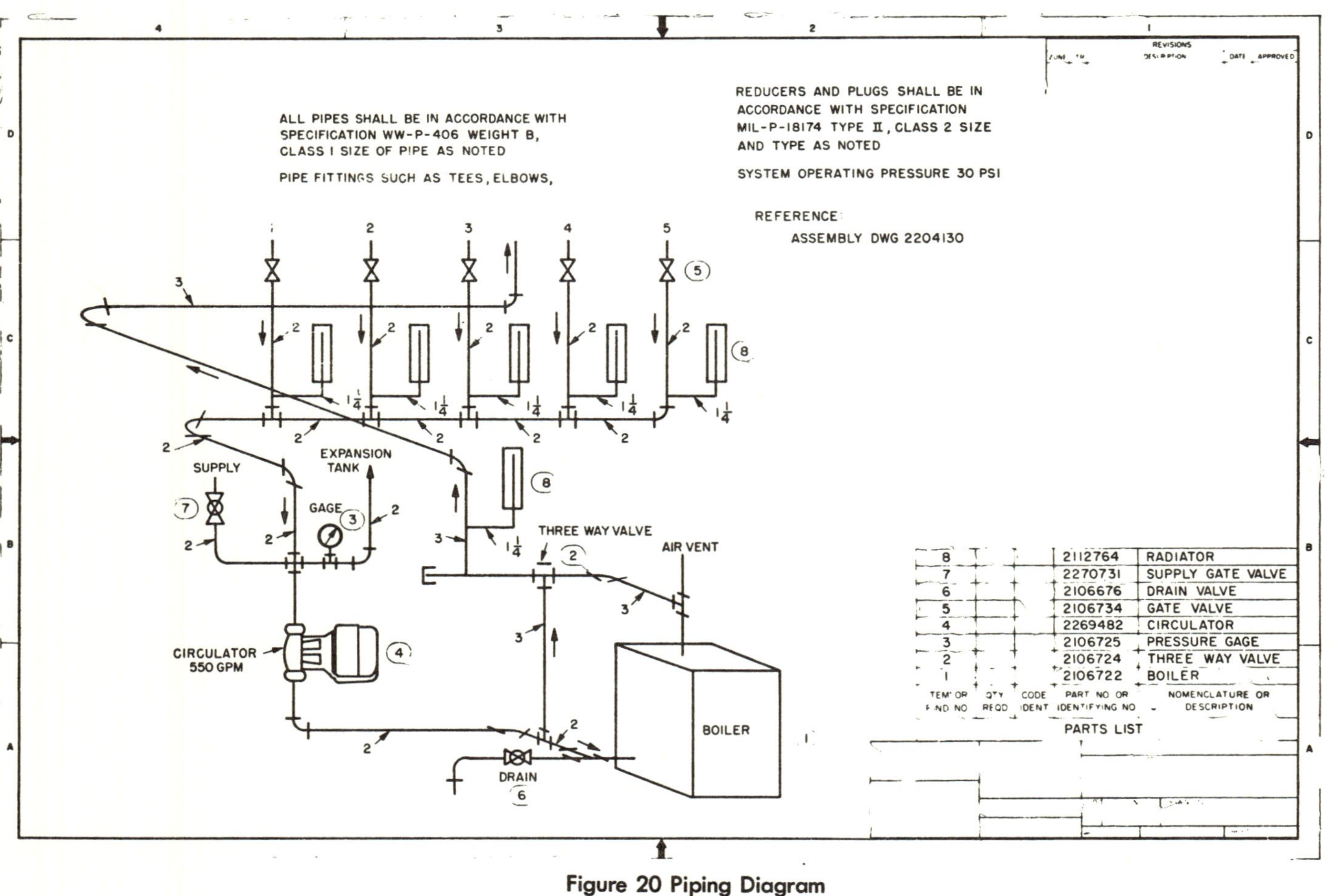

Figure 20 Piping Diagram

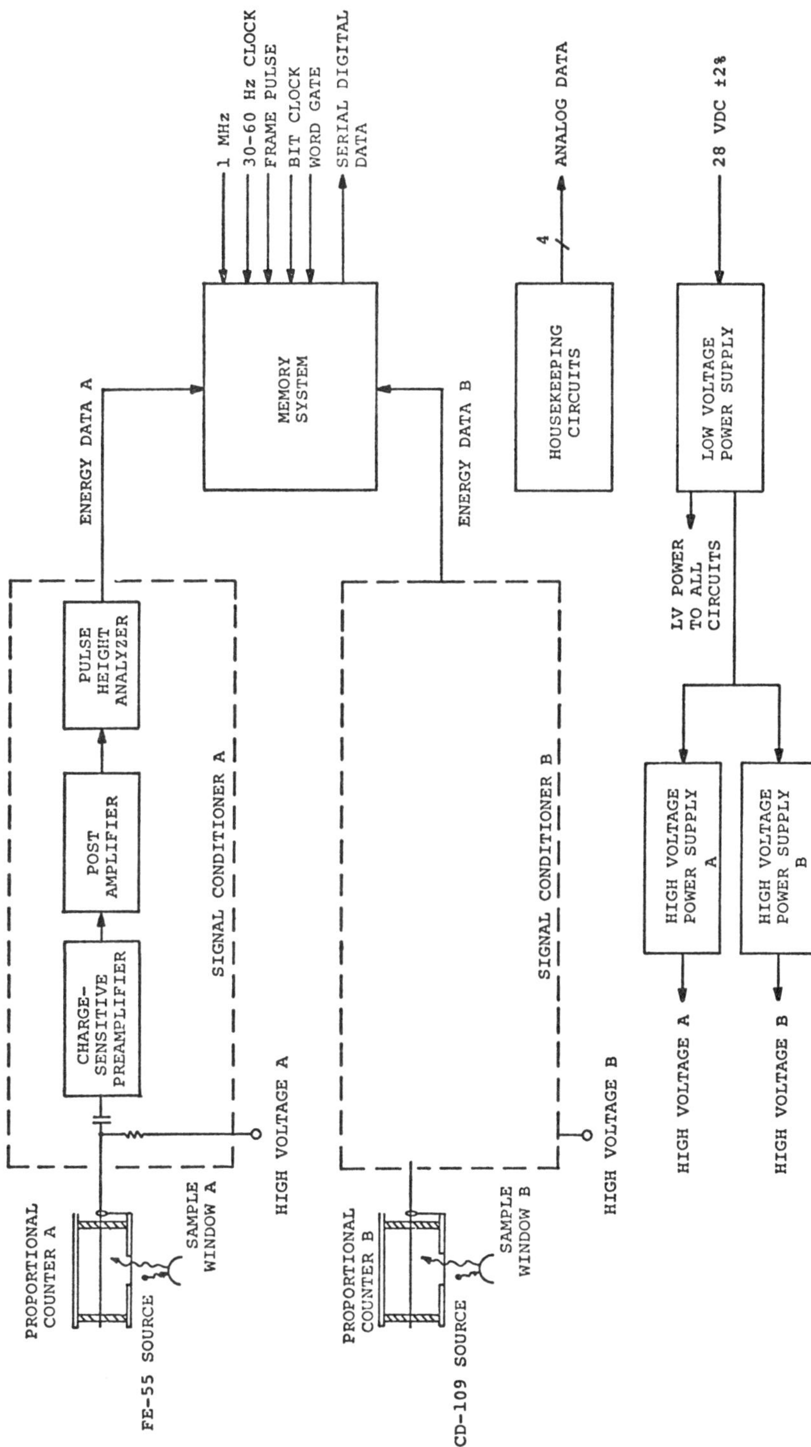

Figure 21 Block Diagram

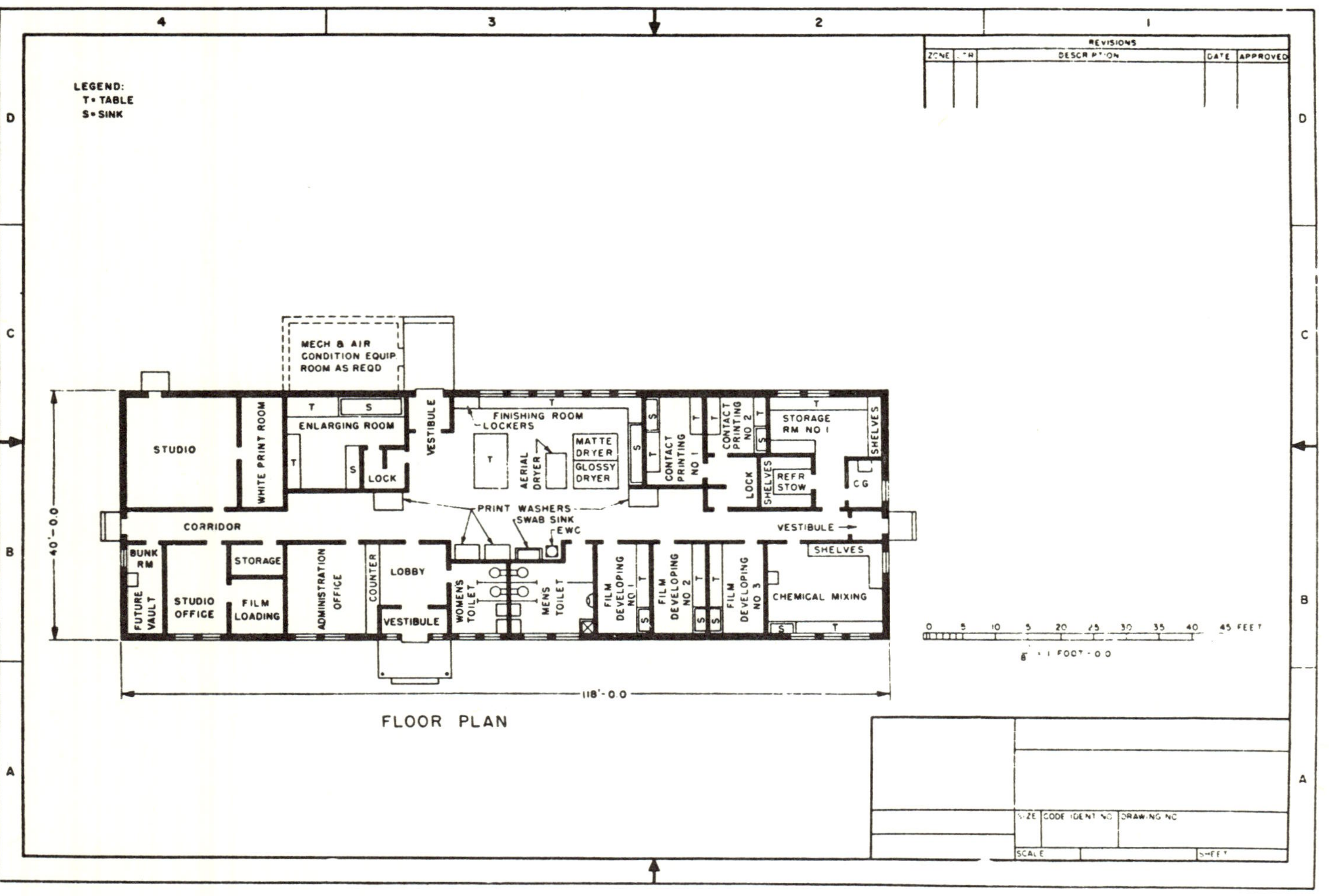

Figure 22 Plan Drawing

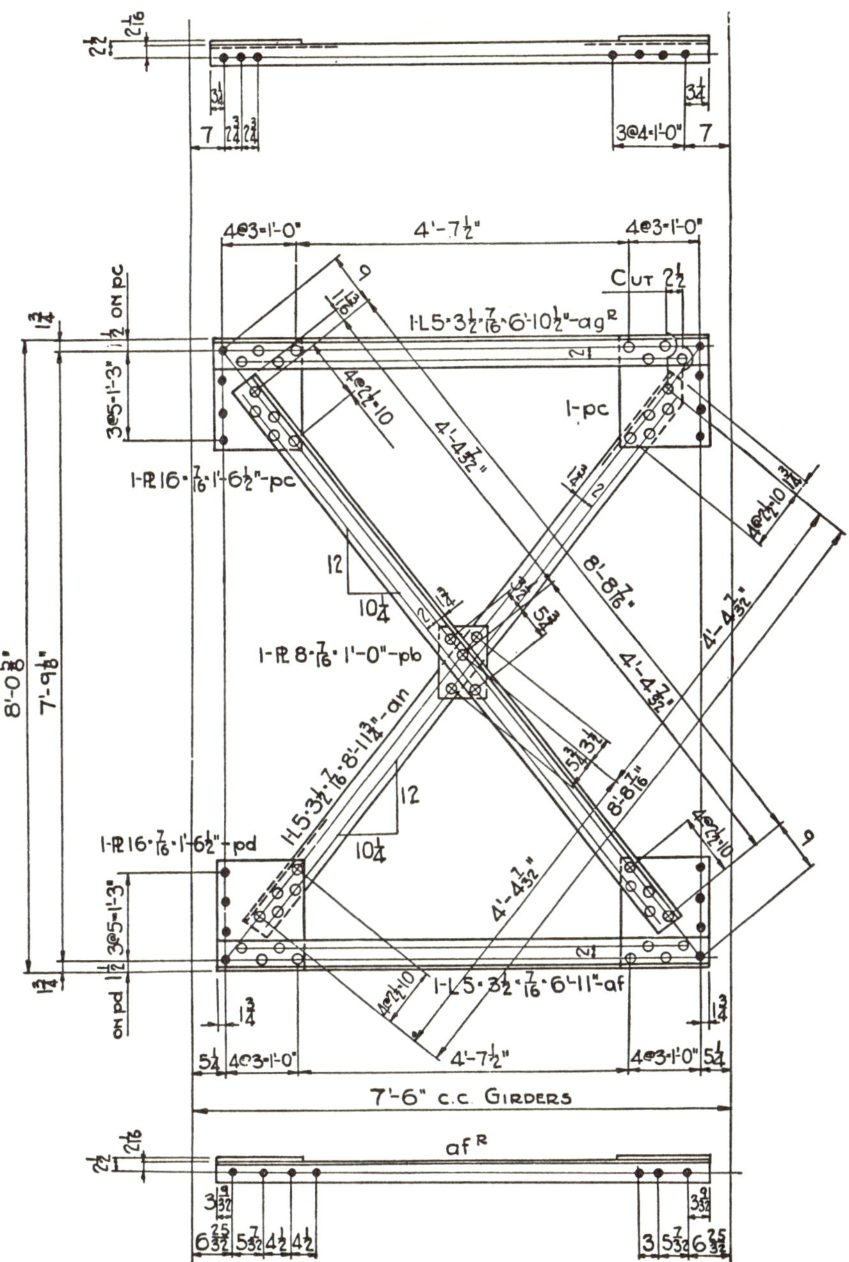

INTERNAL CROSS FRAME FOR
96'-0" DECK PLATE GIRDER
CHESAPEAKE & OHIO RAILWAY CO.

Figure 23. Detail Drawing of Cross Frame

Courtesy of Prentice-Hall, Luzadder, *Fundamentals of Engineering Drawing*, P-H, 1946, p. 491

Assembly drawings form a large class of structural drawings. These drawings include basic and center lines to locate the working points, structural members, gage lines, location dimensions, notes, and other needed data.

Detail drawings give all dimensions and information needed to build or make the parts of a structure. The location and kind of bolts, plates, ring connectors, and rivets are shown with everything drawn to scale.

MAPS AND TOPOGRAPHIC DRAWINGS—A map drawing (see Figure 24) depicts part or all of the earth's surface. When it includes graphical representations of the principal natural and artificial features of the area depicted, it is called a topographic map. The features include lakes, streams, canals, buildings, communication and transportation lines, bridges, fields, meadows, trees, marshes, and other important features. Elevations of several parts of the area may also be shown by contour lines. Special types of maps include:

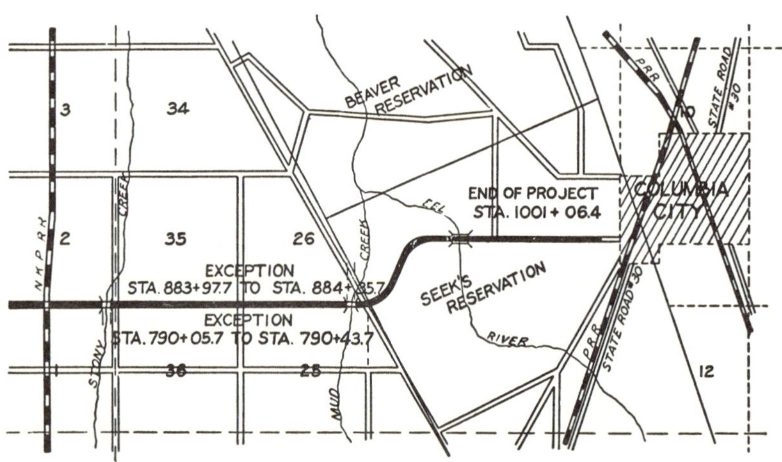

Figure 24 Section of an Engineering Map

Courtesy of Prentice-Hall, Luzadder, *Fundamentals of Engineering Drawing*, P-H, 1946, p. 501

Plats of a survey: record of boundaries of a tract of land, acreage, and identification of this land.

City plat (map of city): used to record street improvements and to specify location of utilities and sizes and locations of property for tax assessments.

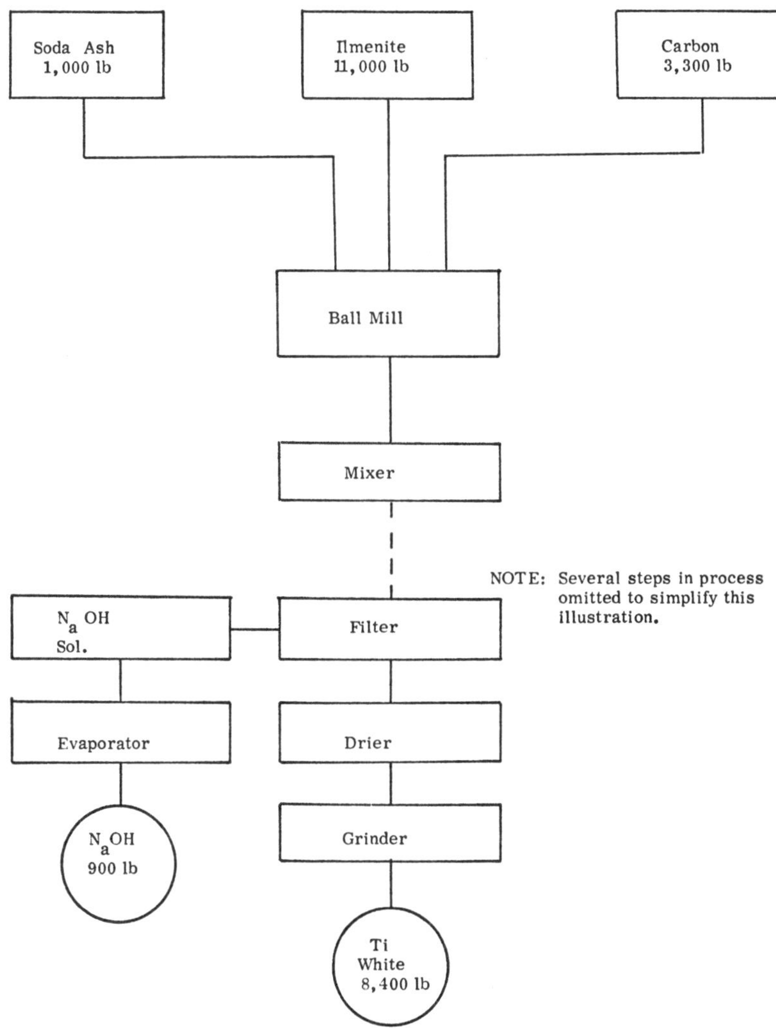

Figure 25 Flow Sheet of Chemical Process

Contours: map illustrating variations in ground levels by use of lines of constant level called contours.

Topographic: give complete description of areas shown. Include boundaries, natural features, works of man, vegetation, and relief (elevations and depressions).

FLOW SHEET—Flow sheets are used to depict various steps in a process or method of operation. Figure 25 shows a typical flow sheet applied to a chemical process. The advantage of this kind of drawing is that it provides information that is readily understood. Without it, the same information would have to be obtained by studying textual material.

MISCELLANEOUS DRAWINGS—Miscellaneous drawings are documents that are used to support basic configuration and fabrication drawings. They include cabling drawings, elevation, printed wire masters, kit drawings, and numerical control drawings. Note, however, that because these drawings are lumped under a miscellaneous category they are not less important than the preceding drawings. These are essential documents and must meet the same high standards for accuracy and completeness as the drawings previously discussed.

CABLE ASSEMBLY—A cable assembly drawing shows the make up of a cable that connects between two or more units. This drawing should contain the following information: length, exact connector or plug types, type of cable or wire, interconnection terminals or pins, labels or marking, and test requirements. See Figure 26.

BENEFITS OF A CABLE ASSEMBLY—A cable assembly allows the manufacture of the right cable for the desired application. Its benefits are (1) cheaper to make cable because the right length cable or wires are specified with proper connections; and (2) allows easier checking of the cable for defects. A cable drawing should be prepared when several or more cables will be made and when a reference document is needed for future work such as checking out the cable for defects.

WIRING HARNESS—A wiring harness shows the arrangement of wires grouped together for convenient and neat connection to their various terminations. The drawing should identify the lengths, colors, materials, and binding techniques. Associated schematics or wire lists are referenced on the drawing.

ELEVATION DRAWING—An elevation drawing is used in architectural work to show the vertical projection or front of a building. However, elevation drawings can also be used to depict vertical profiles of equipment such as ships, automobiles, and aircraft. Figure 27 shows an elevation view of a ship. An elevation drawing shows overall configuration of the object, including the shapes and sizes of its various features. Other information may include walls, compartments, assignment of space, location and arrangement of machinery, and construction materials.

PRINTED CIRCUIT BOARD MASTER ARTWORK—A printed circuit (or wiring) board master artwork depicts the electrical paths of small conductor wires for connection of discrete components between the two sides of the board. It is prepared in much larger size than the final product or board will be, e.g., 4 to 10 times. Multi-views may be presented on one master when

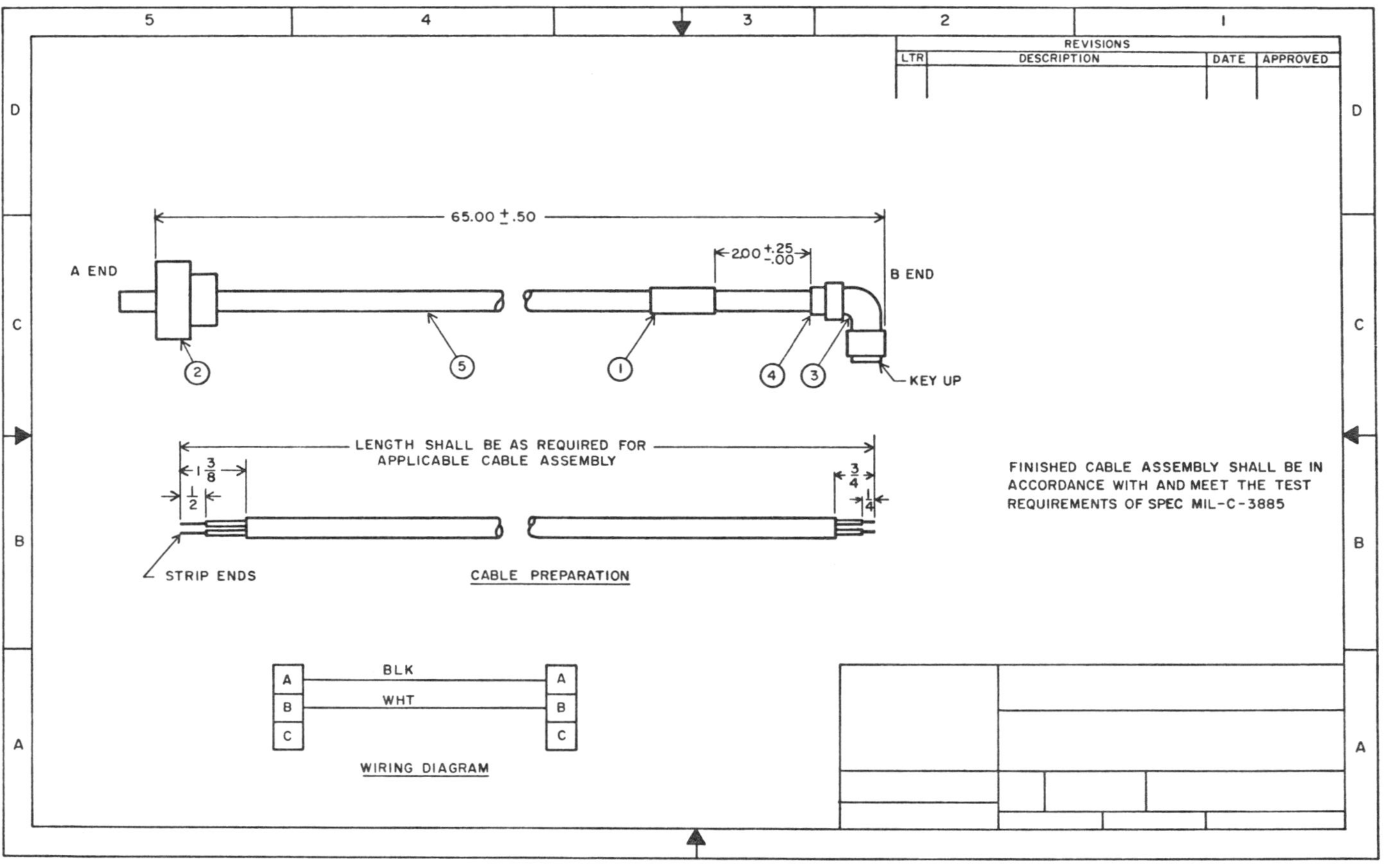

Figure 26 Cable Assembly Drawing

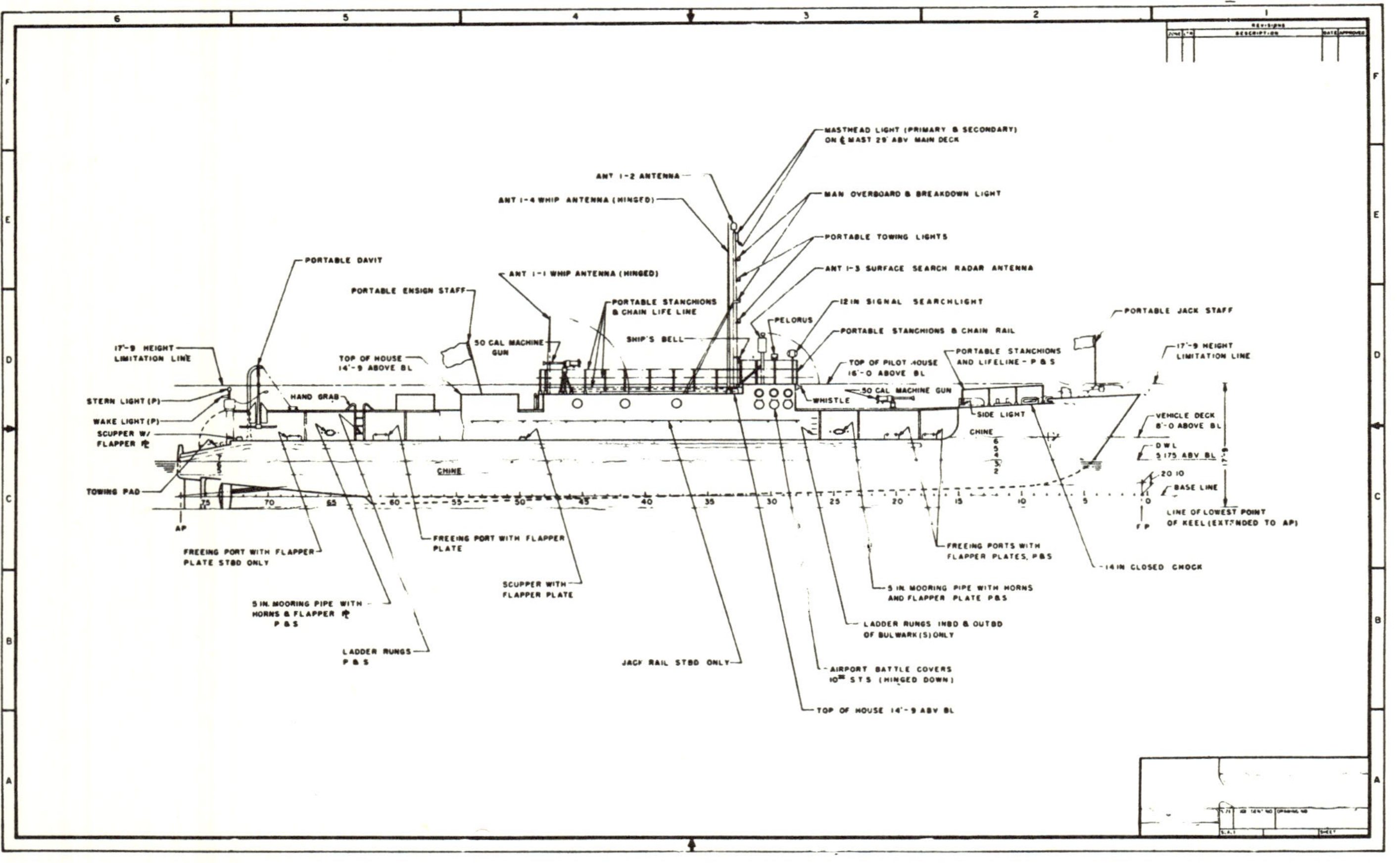

Figure 27 Elevation Drawing

double-sided or multi-layer boards are to be made. Because of the critical nature of this drawing, it is prepared on a dimensionally and thermally stable translucent polyester material. Black, opaque adhesive tape is used to identify the paths of the electrical conductors and component attachment points. In addition, registration marks are added to assure proper alignment of multilayered boards. The final reduction size of the negative to be used for preparing the actual reduced-size board is also given. For example, "REDUCE TO 3.000 ± .005 INCHES."

When the master is completed a contact print is made using an environmentally stable and rugged polyester film. This film is then used to make a reduced negative that is the actual size of the board. This negative is placed over a board coated with conductive material, which is also coated with a protective film. The negative, with the board underneath, is then exposed to light. The light that passes through the clear areas of the negative changes the chemical composition of the protective film. The board is then dipped into an etching solution that removes the metal coating on the board which was not exposed to light. The final board then has a conductor pattern identical to the original taped up master, except that it is the reduced size desired. Sample printed wiring board master negatives are shown in Figure 28.

KIT DRAWING—A kit drawing shows the items required to modify an assembly or product in the field. It includes instructions for modifying the assembly, lists of parts required and their quantities, and sketches or drawings of the modified assembly.

BENEFITS OF KIT DRAWINGS—The value of a kit drawing is in its depiction of a modification package to a device already delivered to a customer. The benefits include: (1) saves labor by identifying all the parts needed to modify the equipment, (2) reduces errors by showing how the change will be made, and (3) provides historical information for future use to allow determination or verification of what was done to the product. A kit drawing should be prepared when the modification is critical and must be carefully defined and controlled. In non-critical modifications, the cost of the kit should be worth the benefits to be derived from formal documentation of the change.

NUMERICAL CONTROL DRAWING—A numerical control drawing gives complete information for the manufacture of the item by a numerical control machine using tape, which contains instructions for making the part; e.g., datum reference points and X and Y coordinate locations for various points on the item to be made are punched into the control tape. (See Figure 29.)

BENEFITS OF NUMERICAL CONTROL DRAWING—A numerical control drawing provides information to a parts programmer for preparation of a numerical control tape for telling the machine how to make the part. Benefits of

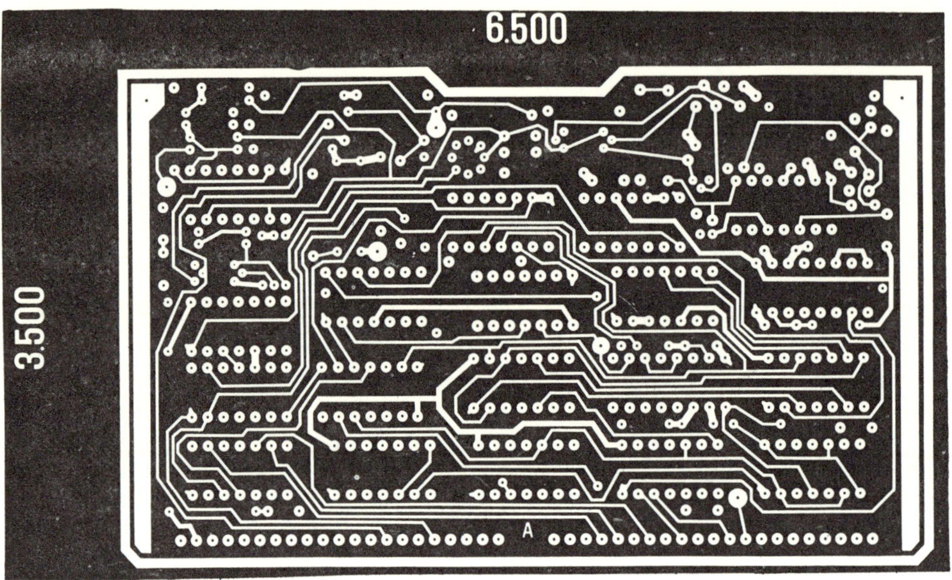

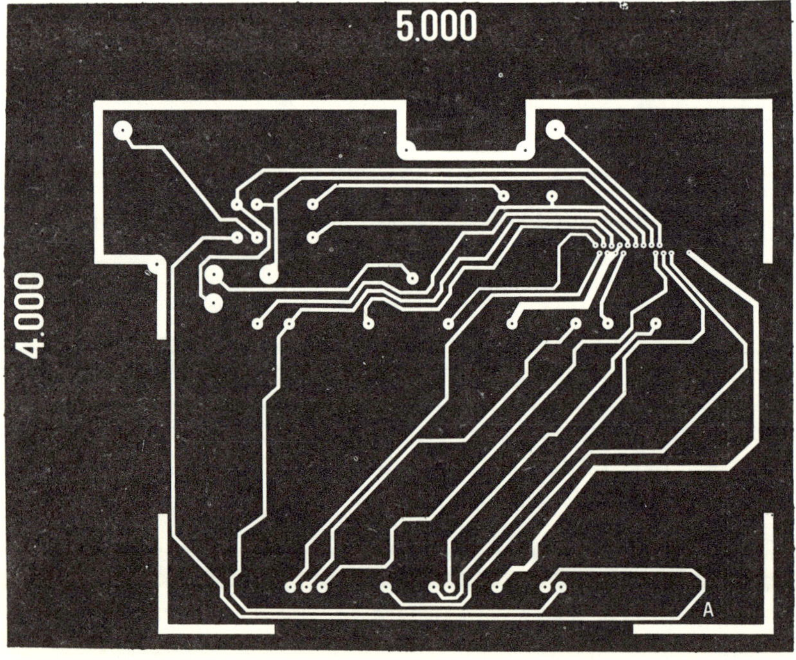

Figure 28 Sample Artwork Negatives for Printed Wiring Boards

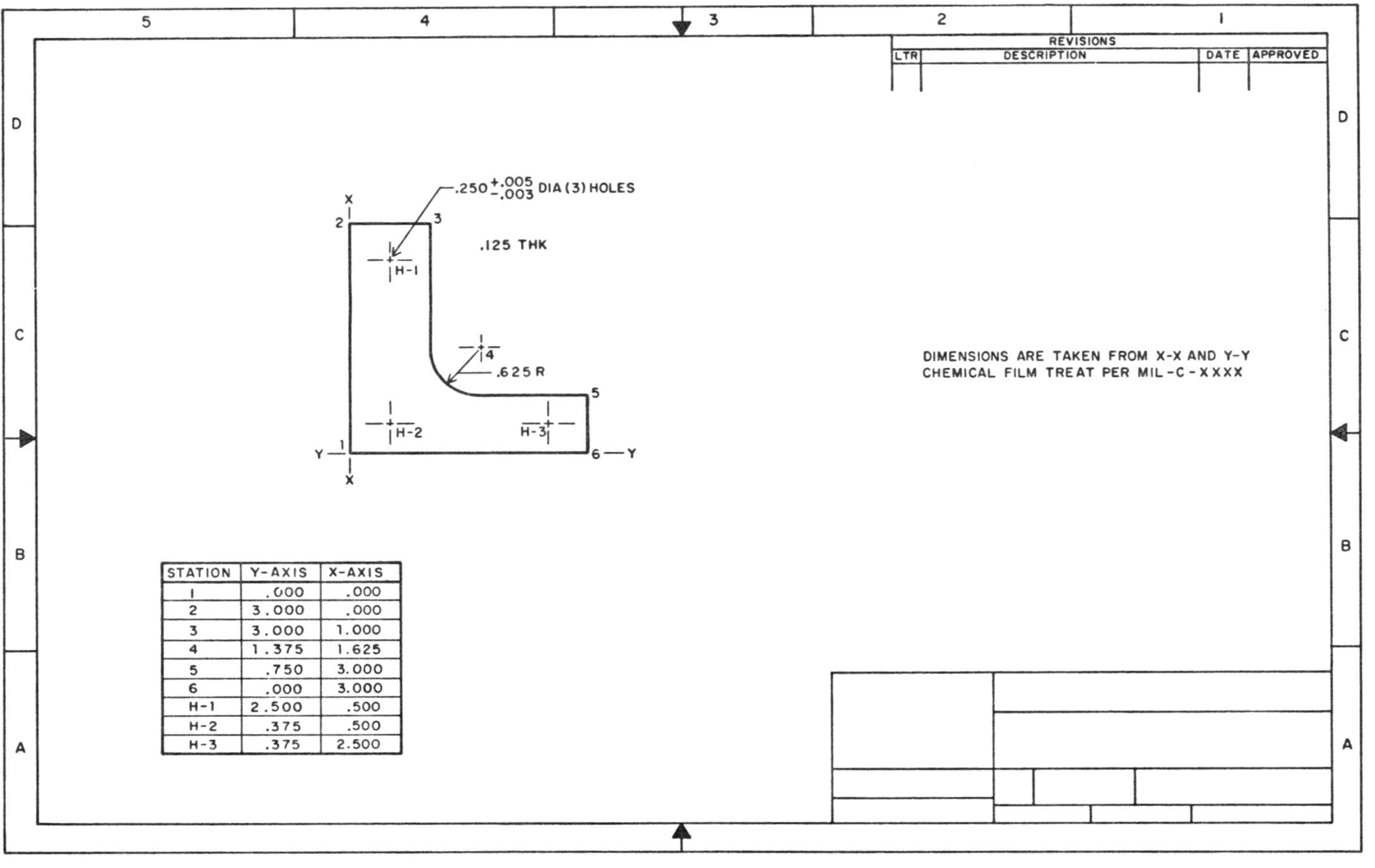

STATION	Y-AXIS	X-AXIS
1	.000	.000
2	3.000	.000
3	3.000	1.000
4	1.375	1.625
5	.750	3.000
6	.000	3.000
H-1	2.500	.500
H-2	.375	.500
H-3	.375	2.500

Figure 29 Numerical Control Drawing

a numerical control drawing are: (1) allows parts programmer to work directly from the print to make the tape more quickly; (2) permits machine operator and inspector to work directly from the print; (3) reduces or eliminates need for calculating coordinates; and (4) provides permanent documentation for future reference or controlled revision.

Numerical control drawings should be prepared whenever a part has been designated for numerical control machine work.

DESIGN LAYOUT DRAWING—A layout drawing shows the original design concept and the placement and spacing of all components required for the item. It is basically an engineering or design work sheet from which detail and assembly drawings are made. It is not released as a production drawing and is identified with the word "LAYOUT" in the title block. This drawing should be made to an accuracy consistent with its ready conversion into a production type drawing.

The layout drawing includes design data requiring special notes by the designer or draftsman; e.g., unique or unusual features, clearances, and critical tolerances. (Layout drawings are delivered with their detail drawings to the checker to enable him to verify the accuracy of the detail design.)

PROPOSAL DRAWING—Proposal drawings show a new design or one involving a major change recommended by an engineer for consideration by management or a customer. Complete data to build the item is not given because it is often unavailable at the time of presentation and because of proprietary considerations. However a weight estimate may be included. Proposal drawings are often not signed by anyone. However, some companies may require drawing approval by the proposal design engineer.

BENEFITS OF PROPOSAL DRAWING—The value of a proposal drawing is in giving the prospective customer a readily understandable representation of the product's configuration. Its benefits are (1) give the potential customer a clear, simplified presentation of the product; (2) increase impact of proposal text; and (3) cost less than a more detailed, complex, and accurately prepared drawing for production use. Proposal drawings are used when a solicited or unsolicited presentation is made to a customer for a system that will meet his needs.

SKETCH DRAWING—A sketch drawing is usually a freehand document on a standard format. It is used when it is necessary to quickly and cheaply supply the shop with fabrication data to build R&D items. Its scale is usually not important to its function and therefore the word "NONE" is placed in the block noted as "SCALE."

Although this drawing should contain complete information, the call out of materials and standard quality control requirements is not usually needed.

The signatures of the draftsman and responsible engineer are needed to release the sketch to manufacturing or the shop.

EXPERIMENTAL DRAWING—An experimental drawing is used to build products in limited quantity; for example, prototype or R&D items. These drawings are prepared on standard drawing format, fully dimensioned with data on manufacturing, assembling, and installing the item. Standard commercial materials can be used and government specifications need not be referenced. The signatures of the project engineer, checker, and draftsman are needed for release.

BENEFITS OF EXPERIMENTAL DRAWINGS—Experimental drawings provide a fast, inexpensive way of allowing skilled people to build early development models of a product. Specific benefits include: (1) cheaper because they are prepared to less stringent drafting requirements and can be changed with less effort and controls; and (2) can be prepared faster because fewer details, notes, and other ancillary data are required compared to full production drawings.

Experimental drawings should be used during the early phase of a program that involves untried methods or materials and requires considerable redefining and redesigning before the final configuration is developed.

PICTORIAL DRAWING—A pictorial drawing shows the external and internal appearance of a product. Pictorial drawings can be isometric, dimetric, trimetric, oblique, or perspective. The main purpose of this type of drawing is to illustrate the overall appearance of the product and is usually not prepared for fabrication, installation, or test purposes. These drawings are not drawn on standard format paper; they usually appear in manuals, catalogs and other types of documents.

BENEFITS OF PICTORIAL DRAWINGS—A pictorial drawing allows a person without technical training to visualize the product presented. It also allows the designer to visualize and develop successive stages of the design. Other benefits include simplicity and ease of understanding and a unique graphical representation of the product not provided by most other types of drawings.

It should be used when it is needed to have an accurate and technically correct drawing that can present the product's configuration as if a person were looking at a three dimensional object. Because a pictorial drawing shows only the appearance of an item, it is not satisfactory for completely describing complex or detailed forms.

4.2 How To Prepare Associated Engineering Data

Drawings often need additional support documents to be used effectively. These documents include parts lists, wire lists, data lists, index lists, test procedures, and specifications. Most of these documents are prepared on standard 8½ by 11 inch format sheets and assembled into book form. However, larger documents are sometimes used. The following sections describe each of these document types.

PARTS LIST (ALSO CALLED BILL OF MATERIALS OR LIST OF MATERIALS)

A parts list contains a tabulation of all materials and parts used in an item. (All assemblies and subassemblies must have either a separate parts list or one integral with the drawing.) Parts are usually listed according to some pattern or grouping. For example parts may be listed by category in the following order:

— Subassemblies
— Detail parts
— Specification or source control parts
— AN, MS, or NAS parts
— Bulk materials (solder, raw stock, bonding compounds, etc.)
— Reference documents (e.g., schematics, wire lists, test procedures, etc.)

Parts may also be listed in their order of need during the production or assembly operation or in alphanumeric order to simplify production and material control operations. Note that although different systems of listing parts may be used, they usually have similar types of parts grouped together. Thus, capacitors would be separately grouped from relays or gears.

Parts and materials are identified as completely as possible. Find or reference numbers for correlating the item with its drawing are necessary. In addition, the quantity required, manufacturer or vendor, and part number must be recorded. A brief description of the item such as "composition resistor, ½ watt, 50000 ohms," should also be entered. When process or other types of control specifications apply, these should be identified. The specification number of the applicable document is recorded in the specification column. The grade or class of the item should also be given.

A symbol column may also be used in a parts list. This column is used to provide additional information on the part in an abbreviated form. For example, "SCD" is used to indicate that a specification or source control drawing number is recorded in the part number column, "B" is used to identify the item as bulk material, and "V" is used to identify a vendor part.

Note that spaces can be left between groups of similar parts listed for later revision without redoing the whole list. This is useful when design changes are frequent during the product's development and early production stages.

Parts lists should contain the same identifying information as their associated drawings such as title, number, and project name. However, the drawing number is prefixed with PL to indicate that it is the parts list for the drawing. See Figure 30 for a sample parts list.

Remember that a parts list is a part of the drawing and is therefore subject to the same format, quality, and approval requirements applied to the drawing. Revisions to parts lists must also be identified and recorded. However, revision of a parts list may not require revision of its drawing; therefore the revision letters will not always match.

Identification of part numbers in the parts list should follow these rules:

- An item described by an approved standard is identified by the standard part number. (Assuming that the item has not been altered or selected.)
- An item described by a government specification containing a part identification system and used without alteration is identified by the specification part number.
- When a part designed and made by another company is used, its original part number is used unless the part is changed or specially selected.
- When a part is purchased to a specification or source control drawing, the SCD number is recorded. It is also referenced in a symbol column, if one is used, or in the specification column as SCD.
- When a repairable item is purchased to a source control drawing from more than one vendor and repairable parts from each vendor are not interchangeable, a dash number is used with the drawing number to identify the source.

WIRE LIST (RUNNING LIST)

A wire list provides the data needed to wire a component, subassembly, or assembly. The wire list is similar to a parts list in format and identification, except that the drawing number is preceded by WL to indicate wire list.

Wire lists contain all the connections to be made on the associated item shown on the drawing. The column headings and entries are shown below.

From component X terminal no.	Wire number	Wire color	To component Y terminal no.	Function
TB 10-1	W1	Black	TB 20-3	30 Vdc
TB 10-2	W2	Blue	TB 20-7	Signal out
......	..	...	...	...
......	..	...	...	...
......	..	...	...	...

PARTS LIST	1. AGENCY (CONTRACTOR IDENTIFICATION):	1A. CONTRACT NO.	2. CODE IDENT	3. PL	4. REVISION LTR DATE
5. LIST TITLE:		6. AUTHENTICATION:		7 REV AUTH NO	8. SHEET OF SHEETS

9. ITEM OR FIND NUMBER	10. QTY RQD	10. QTY RQD	10. QTY RQD	11. UNIT OF MEASURE	12. CODE IDENT.	13. DRAWING OR DOCUMENT NUMBER	14. PART OR IDENTIFYING NUMBER	15. NOMENCLATURE OR DESCRIPTION

Figure 30 Sample Parts List Format

In addition to the preceding, the wire size and type may be added in another column if it differs from wire to wire. If they are all the same except for color, the wire type may be recorded only once on the top of the list or at some other easily visible location.

DATA LISTS

A data list is a tabulation of engineering drawings, specifications, procedures, and other reference documents needed to build, assemble, and test an item. It provides a complete convenient, and current listing of all documents required for the project. Thus, when a document is needed, the data list can be used to find it quickly. For projects not using many documents other than drawings, a drawing list may be used instead.

The identification format is similar to other lists. The drawing number for the top assembly or end item can be used with the prefix DL to indicate data list. Columns in the list can include the item number, document number (latest revision letter), number of sheets, drawing size, and title of the document. For government contracts, a code identification number column is also needed. Data lists should be approved before release or revision.

Listings of documents should be grouped by category and in the order shown below:

Drawings
Specifications and standards
Government drawings
Government specifications and standards
Vendor data

INDEX LIST

An index list is a tabulation of data lists for a system. Its purpose is to provide a complete identification and collection of data lists prepared by various projects contributing to a larger system effort. This list is not required for a small project or program. Data list information includes the following columns:

Identification number of data list
Revision letter of data list
Title of data list
Name of project or subsystem
Code identification number of company originating data list

DATA SUBMITTAL—Data submittals involve collecting all the engineering data prepared for a program, preparing data lists and EDP cards, and producing duplicate vellums or microfilm from the original documents. When done to stringent customer requirements, considerable effort and cost results. Costs can exceed thousands of dollars. A typical data submittal includes: (1) production drawings; (2) associated parts and wire lists; (3) reference documents; (4) test procedures/specifications; (5) specification control drawings; (6) data lists; (7) artwork for printed wiring boards; and (8) electronic data processing cards. This data package is submitted to the customer for reprocurement of the product from another manufacturer or for future reference.

INDENTURED DRAWING LIST AND SCHEDULE

An indentured drawing list identifies all the items that make up the end item. It contains a tabulation of all the drawings in terms of their subordinate drawings. Thus, an assembly drawing XXXXXX will have its subassembly and component drawings indexed to show that they go into the higher level assembly XXXXXX. The higher level assembly in turn is indexed to show that it goes into the next higher assembly.

Indexing can be done by listing a series of numbers in a column next to the drawing number that indicate the level of indenture by their magnitude. For instance, 1 can be used to indicate the top assembly, 2 the next assembly below it, and 3 the next assembly below the 2 level. This process can continue until all subordinate or lower levels of assembly have been identified. Another method of indexing is to physically position the drawing number to the right of its next higher assembly. Equal level assemblies retain alignment with each other. Thus:

123450 .... highest level of assembly
  123451 .... goes into 123450 assembly
  123452 .... goes into 123450; same level as 123451
    123455 .... goes into 123452
      123459 .... goes into 123455

The previously mentioned method of using numbers to show indenture level is shown next.* As mentioned, the degree of indenture depends on the product complexity and level of detailing required.

Indentured Level

1	2	3	4	5	Drawing number	Title
X					123450	Top assembly for product
	X				123455	Subassembly 1; goes into 123450
		X			123460	Component 1 of subassembly 1
		X			123475	Component 2 of subassembly 1

*Adapted from Samaras and Czerwinski, *Fundamentals of Configuration Management*, Wiley-Interscience, New York, 1971.

	X	123479	Module; goes into Component 2
X		123480	Subassembly 2; goes into 123450

In addition to the drawing list, a schedule can be added to the same pages to indicate the status of each drawing; e.g., in work, in review, in check, or released to production. Although the combined drawing list and schedule is a useful document, many companies do not use it. Thus, the drawing list may be a separate document. In this event, the following columns should be used:

Item number
Drawing number
Drawing title
Revision letter
Drawing size
Assembly level (indenture number)

ASSEMBLY AND TEST PROCEDURES

Assembly and test procedures are often required to assemble and checkout a part, subassembly, etc. These documents are usually referenced on the drawing by their identification numbers and titles.

Whether assembly or test, these documents contain similar outlines and headings. For example, a typical test procedure includes the following:

1. Purpose of test
2. Applicable or reference documents
3. Test requirements
4. Test fixtures and equipment
5. Test setup
6. Test instructions
7. Data sheets

An assembly procedure could include the following:

1. Purpose of item
2. Applicable or reference documents (drawings, process specifications, etc.)
3. Assembly requirements
4. Assembly equipment, tools, and materials
5. Assembly setup (drawing)
6. Assembly instruction

For many standard assembly procedures, written instructions may not be needed. But some companies prepare assembly or abbreviated manufacturing instructions even when the procedure is well-known and experienced production personnel are used. This is done when high quality workmanship is demanded by management or random variations in procedures reduce efficiency and lower product integrity and reliability. Written procedures also reduce errors and allow inspection points to be built into the system.

SPECIFICATIONS: KEYS TO PRODUCT DEFINITION AND CONTROL

Specifications are documents that define the requirements for designing, manufacturing, processing, and testing items or products. They also include methods for determining that the requirements have been met. Specifications fall into two basic categories: performance and design. A performance specification gives requirements for an item in terms of output, function, or operation. The details of design, fabrication, and internal operation are left to the supplier's judgment. A design specification on the other hand is more definitive. The requirements in this document are given in sufficient detail to reproduce the item exactly. For example, detailed information on materials, components, weight, size, dimensions, and shape is given.

Specifications are usually referenced on the drawing or parts list and should be carefully checked by the draftsman to ensure correct identification. A few types of specifications are described next.

Industrial Commodity Specifications

Material Specification: gives properties and detailed requirements for a raw, semifabricated, or processed material supplied for additional processing or use in a finished product.

Part Specification: establishes design and detailed manufacturing, inspection, and test requirements for an item or part.

Equipment Specification: gives functional, performance, and physical requirements for an end item or combination of items that perform a specific function. Similar specifications exist for subsystems and systems.

Industrial Process Specifications

Manufacturing Process Specification: establishes material properties and detailed process control requirements for materials or items that require specific manufacturing operations such as welding, heat treating, potting.

Finish Specification: gives the method and requirements for protective treatments and finishes for materials, parts, and other items.

Packaging Specification: specifies the method and detailed requirements for packaging, handling, preserving, storing, or shipping an item.

Marking/Identification Specification: gives method and requirements for identifying an item or shipping container.

Interface Specification: establishes requirements for mating two or more items.

Environmental Test Specification: gives environmental test and performance evaluation requirements (temperature, humidity, vibration, and shock) for qualification or acceptance testing.

Architectural Specifications

Architectural specifications give detailed accounts of material and workmanship for various structural parts. These specifications also give requirements for site work, doors and windows, finishes, and electrical fixtures and wiring. Architectural specifications are similar to industrial specifications but have their own specialized features.

4.3 Cost and Performance Considerations

Cost reduction and performance improvement efforts in the areas covered by this chapter involve the use of smaller size drawings, existing engineering data, computerized parts lists, and firm guidelines defining format and content of various types of drawings and associated data. In addition, photographic techniques, drafting aids, functional drafting, and typing on drawings are methods for reducing D&D costs. These techniques are described next.

SMALLER DRAWINGS

Smaller drawings should be used whenever possible if their use doesn't increase costs in other areas such as engineering, production, and inspection. Smaller drawings can provide cost savings in the following areas:

- Smaller, cheaper drafting tables can be used.
- Use of smaller drafting tables saves space and thereby reduces rental and other overhead costs.
- Data filing space is reduced. As a result fewer file cabinets and less floor space are needed.
- Lower material (e.g., paper) costs result.
- Labor costs are reduced because production personnel can work more efficiently with smaller drawings and reproduction workers can make prints and fold them faster.
- Need for intermediate vellums may be eliminated if the drawings are small enough to be copied by xerographic equipment.

The last item is obtainable using a xerographic reduction machine that automatically makes and folds copies in seconds. The machine converts up to D size drawings to folded B size. The labor savings here can be considerable if print requirements are large.

EXISTING ENGINEERING DATA

Costs can be reduced by using existing data in exact form or through modifying it. This includes drawings, parts lists, wire lists, specifications, and procedures. To be effectively used, D&D personnel should have a convenient and fast system for reviewing existing data for use on a new project, or product.

D&D personnel should also be thoroughly familiar with existing standard parts documents such as specification control drawings, so that they can be used when applicable.

STANDARD PARTS DRAWINGS & CATALOGS

When a part is used over and over again, it can be made into a standard part by preparing a control drawing that establishes uniform requirements simplifying purchase. Instead of redefining past specifications each time the part is ordered on a purchase order, a copy of the control drawing is attached to the purchase order, thereby reducing purchasing, design, engineering, and inspection labor. Standard part drawings should be compiled into a single catalog for easy reference and updating.

WELL-DEFINED FORMATS AND CONTENTS

Establish firm guidelines for preparing various types of drawings and engineering data to avoid rework and omissions. Have sample drawings of each type in the D&D manual or a special notebook. Make sure everyone in D&D knows that they are available and where to find them. Use preprinted forms whenever a format is used frequently.

SIMPLICITY AND FUNCTIONALITY

Drawings should be prepared containing only lines, views, notes, etc., required for clear and unambiguous understanding of the subject by the user. Extra lines, dimensions, shading, cross hatching, and useless details should not be used when they serve no purpose.

The primary objective of D&D personnel should always be functionality.

To achieve this objective, a review of drafting practices should be made to detect unnecessary drafting work.

Be sure that detailed views of a part or component are not repeated on some other drawing for the project. If an item is shown elsewhere it should not be repeated on another drawing unless other considerations apply such as ease of fabrication, assembly, or testing. Some guidelines for minimizing the complexity of drawings follow:

1. Use a minimum number of principal views.
2. Rearrange principal views to eliminate or reduce auxiliary views.
3. Use partial views and broken-out sections whenever feasible.
4. Use dotted lines and shading/cross hatching only when they serve a useful function.
5. Eliminate unnecessary details and views of items that can be clearly understood by their outlines or by simple presentations. For example, threads, bolts, and spring coils need not be shown in detail unless for a purpose.
6. Don't use isometric and perspective views unless required.
7. Prepare additional drawings of a simpler nature rather than fewer drawings of a complex configuration.
8. Choose a drawing size that is somewhat larger than needed. This avoids trying to squeeze the drawing onto the vellum and allows additional notes or views to be added in the future without redoing the entire drawing.
9. Use one view drawings when they completely define the item.
10. When a feature occurs many times in a regular pattern, use only enough information to illustrate the feature and its regularity. Thus, a row of holes, bolts, rivet heads, and slots should not be detailed completely.
11. Use descriptions in list of materials to eliminate drawing certain simple parts such as bolts, nuts, gaskets, and tubes.
12. Use descriptions to eliminate projected views.
13. Eliminate unnecessary notes.
14. Avoid hand lettering or reduce to minimum.
15. Avoid use of arrowheads. (If users agree.)
16. Make free-hand sketches when possible.
17. Use ordinate dimensions. With this technique, dimensions are measured from several base surfaces or center lines thereby reducing conventional dimensions and extension lines that tend to clutter the drawing.
18. Take advantage of symmetry to reduce drawing time and size.
19. Use symbols to indicate various size holes.
20. Avoid scaling the drawing if not needed for clarity.
21. Use abbreviations when this practice does not detract from drawing clarity.

PHOTOGRAPHIC AND OTHER TECHNIQUES

The use of photographs to depict a complex item can save many hours of drafting labor. For example, a sophisticated electronic assembly can be photographed instead of being hand drawn. A half-tone image can then be produced on a standard format and call-outs, additional views, and notes added to the vellum by hand. If the drawings are being prepared for a customer, be sure that this kind of presentation is acceptable to him.

Another technique that can save money is to use special wash-off films for restoring old drawings or revising originals/vellums. In the first case, a diazo machine can be used to make a new vellum, which is relatively free of the discoloration, cracks, and creases of the old original. In the second case, the vellum film can be easily altered to accommodate design changes or drafting errors by using a moist eraser to remove the old lines or letters.

APPLIQUÉS AND DRY TRANSFERS

Appliqués are drafting symbol cutouts attached to a waxed sheet of back-up paper. These symbols have adhesive compounds on one side allowing them to be quickly added to a master to save drawing the symbol. If the symbol is placed incorrectly, it can be easily removed. Burnishing is used for permanent bonding.

CASE IN POINT: Oberg Manufacturing Co. reduced drafting time in the development of detail drawings by as much as 50% by using appliqué techniques. This reduction in labor is obtained by using elements of previously prepared design drawings. Dimensionally stable duplicates of these design drawings are made. The desired details are then cut up and placed on a new vellum master that forms the basis for the detail drawings. Only changes and additions are drawn on the new vellums, which are duplicated on vellum to serve as reproduction intermediates.[12]

Dry transfer symbols are available to save routine drafting work. Title blocks, letters logos, logic elements, pads, and components are available in standard or custom dry transfers. Patterns or symbols are transferred from the master sheet to the original document by rubbing the desired symbol with a dull pencil, ball point pen, or burnisher. Special dry transfers (or appliqués) of standard company parts may also be created with special equipment.

OVERLAYS

Overlays can be used to save time and labor. An overlay is a sheet of transparent or translucent film containing printed or graphical data. This sheet is used to add information to drawings without additional drafting labor. For example, an overlay containing a commonly used set of notes can be placed over the original drawing and a new vellum made containing the notes, or an overlay with a special notice (not wanted on the original) may be used when printing copies of certain drawings for distribution to an outside vendor or a customer. An overlay may also be used for title blocks or to block out certain data on the original so that prints will not have it; without the overlay, the information would have to be deleted from the drawing, requiring an expensive engineering order and drafting labor. Note, however, that overlays containing standard notes and parts that are used frequently can be replaced with appliqués.

SHADING SHEETS FOR DRAWINGS

Although shading and cross hatching should be avoided, some drawings require shading certain portions of the item for improved clarity or artistic value (such as in brochures or advertisements). Manual shading can be time consuming and expensive. This work can be avoided by using acetate shading sheets, which are available in various patterns. They have adhesive on one side so that they can be applied directly to the drawing vellum. After the sheet has been gently pressed into position, it is outlined with a cutting needle along the area to be covered by the shading and undesired shading removed. The remaining shading is rubbed down firmly with a burnisher.

TEMPLATES

Templates can reduce drafting time by offering standard outlines to simplify and accelerate the drawing of components, squares, ellipses, circles, and rectangles. Templates are also available for quickly drawing symmetrical dotted and dashed lines when needed. Because templates are relatively cheap and durable, they provide a reliable way of cutting costs when their symbols are needed by draftsmen and designers.

CASE IN POINT: Templates can save much time as well as improve consistency in use and style of symbols put on drawings. For example, a reduction of 40 percent or more can be obtained compared to using a rule, triangle, and compass to

create standard symbols on drawings. In addition, expensive redraw time to revise symbols not made correctly can add to drafting costs.

CUT AND PASTE

When a part of one drawing can be used essentially as is on a new drawing, make a clear high quality reproduction vellum and cut out the desired part. Then the cut out is taped or pasted onto the drawing. Note, however, that the labor in obtaining the other drawing, reproducing it, cutting out the desired item, and taping it to the new drawing vellum must be less than the time required to redraw it. If it is a border line savings, it is usually better to redraw it because of the costs of the materials needed and the chances for waste.

CASE IN POINT: The use of composition (cut and paste) drafting, which eliminates the need for repetitive drafting, is saving up to 80 percent in labor costs at the Mechanical Division of Borg-Warner Corporation. By using the composition drafting system, a drawing can be revised or created and blueprints made in less than a half an hour. This involves cutting out usable components or parts from old drawings and pasting them onto a new original. Consequently, a group of four draftsmen is able to revise and redraw 10,000 old assembly drawings a year while also keeping up with its normal department workload.[13]

AVOID EXCESSIVELY SMALL SCALES

Don't try to draw a complex item using an excessively small scale because it requires greater effort by the draftsman to properly draw the lines required to show the item. Also the chances of errors are increased and the draftsman will suffer from greater fatigue. Of course, if a maximum size drawing has been standardized to simplify reproduction, folding, storage, and handling, the cost savings here may offset the extra work that the draftsman may have to do to get a complex item on the maximum allowable vellum size.

MAINTAIN HIGH MICROFILM QUALITY

If drawings will be microfilmed, they must be done more carefully to ensure acceptable reductions and blow backs. If adequate care is not given to uniform line densities, proper spacing between lines and letters, and adequate size lettering, the drawings will be returned for redrafting or correction, thus duplicating work and delaying progress.

Cleanliness is also important. Incomplete erasures, dirt, smudges, stains, and discolorations due to chemical eradicators will result in poor microfilm copies of the original. See Chapter 7 for more information on microfilming.

USE EXISTING DRAWING STANDARDS

Whenever practical, use existing industrial or government standards for symbols, drafting standards, abbreviations, and dimensioning and tolerance practices. No need, in general, to duplicate work that has already been done and published for guidelines. Obtain a copy of the American National Standards Institute catalog for various standards available. For military and government standards refer to J. W. Parker, DRAWING REQUIREMENTS MANUAL, Global Engineering Documentation Services, Newport Beach, Ca.

STANDARD DRAWINGS

Tabular type drawings present a possible danger area in that the user may misread the table of listed variations. To avoid this problem, a standard drawing may be prepared illustrating the basic part configuration and dimensions that are the same for each variation. This standard drawing is then duplicated in quantities to satisfy the requirements of all the anticipated variations. These duplicate vellums are then identified with their unique document numbers, dimensions, or other variable characteristics for each different configuration. Thus, only variance data need be added without redrawing the part completely for each special case.

TYPING ON DRAWINGS

Typing on drawings can save time, provide more useful application of a draftsman's skill, and provide improved and consistent drawing quality. The draftsman's time can be reduced by eliminating the routine need to hand letter notes, parts lists, and drawing lists. In some cases, a less skilled person can be used to do the typing.

> **CASE IN POINT: At Burroughs, it was found that typing on engineering documents by typists rather than using draftsmen reduced an equivalent of 270 manhours needed for handlettering to 67½ hours by machine. The system uses a Varityper, one-time carbon ribbons, and polyester film for the original drawing. This process eliminated the need to have skilled draftsmen hand letter notes, specifications, parts lists, data lists, wire lists, instruction sheets, and schematic**

diagrams. Other benefits of typing on a drawing were improved and consistent quality and time savings.[14]

Varitypers, or other similar devices, are used for this application because they can handle any size drawing. The Varityper can also have its impression force adjusted so that uniform letters are produced regardless of the force applied to the keys. Interchangeable fonts are also available that can be exchanged in seconds.

Besides the Varityper, drawing board typewriters are available that will type on all drafting surfaces, including film. These typewriters can move in any direction and are quite small, e.g., one device is 3 by 6 by 7 inches and costs about $300.

DRAWING WITH INK

Drawing with ink can be cost effective, especially when drawings are to be microfilmed. Some companies have found that ink drafting can be cheaper than pencil drafting.

CASE IN POINT: The East Bay Municipal Utility District, Oakland, Ca. found ink drafting on film was 10 to 15% cheaper than good pencil work on conventional vellum. This reduced labor was obtained by using polyester film, Rapidograph-type pens, and ultrasonic pen cleaners. Use of a line softener (alcohol) allowed the image to be quickly erased without injury to the original.[15]

COMPUTERIZED PARTS LISTS

Computerized parts lists can provide labor and time savings when the quantity and use of data required is sufficient to offset initial setup costs and the system is designed for maximum utility by the staff. (A computerized parts list, however, should not be used when a good typist and typewriter will do the same job more cheaply and efficiently.) Therefore, to avoid this situation, the system should be designed so that other departments such as estimating, manufacturing, and material control, can obtain useful data when needed. This means that as much data as possible about parts should be added into the computer memory or auxiliary storage so that a greater variety of output reports can be obtained. In addition, programs should be prepared so that output data can be printed out in different formats required by users. For example, an efficient computerized parts list system could do the following for different company users:

User	Report
Engineering	Complete parts list of each assembly, subassembly, and component used in product upon request. It can also contain delivery status of needed parts.
Procurement	Total quantity of a particular part needed in various subassemblies of a product or all products.
Estimating	Prices of various parts used in product.
Scheduling	Ordering leadtimes required for various parts and components.
Manufacturing	Indentured parts listing of all items used in assembly. Also status or location of each item.

Some other advantages of a computerized parts list system are:

1. One central source is used to supply data. Also corrections need to be made to only one data source.
2. Corrected parts lists can be produced very quickly after revised data have been placed into the computer memory.
3. More compact means of storing data.
4. Can be programmed to provide rapid answers to questions that would take many hours by manual means. For example, the part and serial number location of a particular item of doubtful reliability in a large hardware system can be traced down quickly by the computer.

The procedure for inputing data into a computer that has been programmed to process PL data follows:

1. Engineer prepares sketch of item and submits it to D&D.
2. D&D converts data on sketch into a hand-written parts list containing all needed data elements. (A special coding form is used for this purpose.)
3. Handwritten parts list is submitted to the computer facility with a work request.
4. A keypunch operator makes a punched card for each part on the list and the cards are fed into the computer.
5. A checkprint of the PL produced by the computer is returned to D&D for checking against the original PL data inputs.
6. Corrections are fed back to the keypunch operator and the required number of copies of the output report are produced.

When changes are needed, these are processed through a single control point such as data control, which is responsible for controlling change inputs and notifying users of the PL and related data that changes have been made. Whenever the PL is revised, a notice of revision should be issued to the staff specifying the exact nature of each change. (A PL distribution list must be maintained to assure that everyone using the data gets a revision notice.)

Note that it is extremely important to tightly control changes to the stored data just as to released production drawings. In a centralized computer facility, which is restricted to operation by only authorized personnel, this may not be difficult. However, if a time-shared system is used where a central computer is located at some remote location and several input and output terminals are available to the staff, the possibility of uncontrolled changes to the PL data file is much greater.

5

5.0 Identification of Documents and Parts: Key to Configuration Control and Tracking

Identification of documents and the parts made from them is a key responsibility of the D&D department. However, the department is not responsible for actual marking or identification of items but for describing where and how they should be identified.

Incorrect or deficient assignment and marking of specifications, drawings, parts lists, subassemblies, assemblies, and end items leads to chaos in manufacture, test, use, and repair operations. Therefore, D&D personnel must have a sound understanding of the identification principles involved and the systems and procedures to be followed to ensure accurate and complete identification of the company's documentation and hardware.

BENEFITS AND APPLICABILITY OF THIS CHAPTER

Numbering systems exist in all organizations preparing engineering data. Therefore, engineers, designers, draftsmen, and managers must be familiar with their use, construction, and modification. The prime benefit of this chapter is a broad review of numbering systems and how they are used. Specific benefits include:

- Provides advantages and disadvantages of significant and non-significant numbering systems.
- Gives criteria for selecting numbering systems.
- Provides samples of various types of numbering systems.
- Provides review of different part numbering techniques.
- Describes various types of number systems for configuration identification.

- Provides a review of numbering system characteristics and limitations.
- Describes criteria for identifying non-interchangeable parts.

Because all companies have a need for numbering systems, this chapter applies to all companies and organizations, regardless of size.

The following sections are devoted to a review of the various aspects of a documentation and hardware identification system. Additional information can be obtained from articles published in *Engineering Graphics* and *Reprographics*, or from the American Institute of Design and Drafting. The book, *Fundamentals of Configuration Management*, Wiley-Interscience, also has material on product and documentation identification.

5.1 Numbering Systems: Significant Vs. Non-Significant

There are two broad categories of numbering systems. The first is called a significant numbering system; the second is designated as a nonsignificant system. A significant system uses alphanumeric symbols to represent specific kinds of information. For example, a significant drawing and part number such as C11-500T02-10 could tell the user the following without having to look at the drawing:

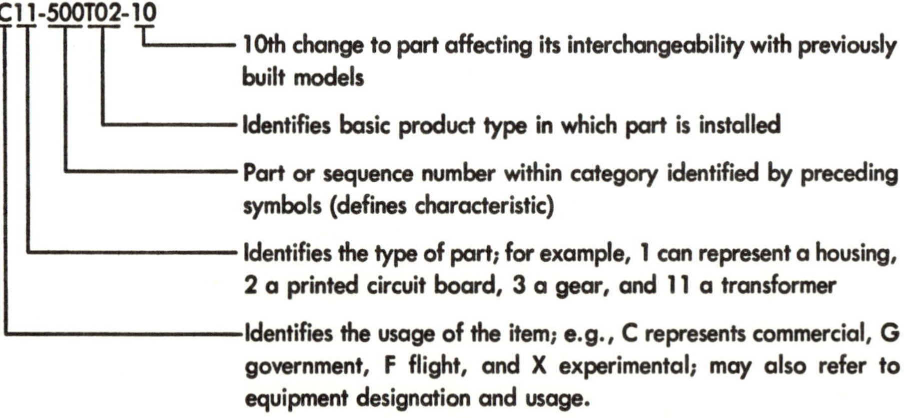

The codes given above are, of course, completely arbitrary in most systems and are established by each company for its own purposes. However, once a coding system is identified, it must be recorded, distributed to everyone affected, and changes controlled by the department responsible for the system. Without centralized control of a company's numbering system, several numbering systems can evolve that will create confusion and errors in interpretation that can be catastrophic.

A nonsignificant numbering system is one that uses unique identifiers that have no coded information. For example, the drawing number 30000 identifies a particular document but does not tell anything about the type of part depicted on the drawing. That is, 30000 is simply the thirty thousandth number issued in sequence to a drawing by data control or some other department that issues drawing numbers in the company.

Of course, numbering systems can fall somewhere in between completely significant and nonsignificant numbering systems. The number 301010 can be coded to provide the user with some information about the drawing identified by this number as shown below:

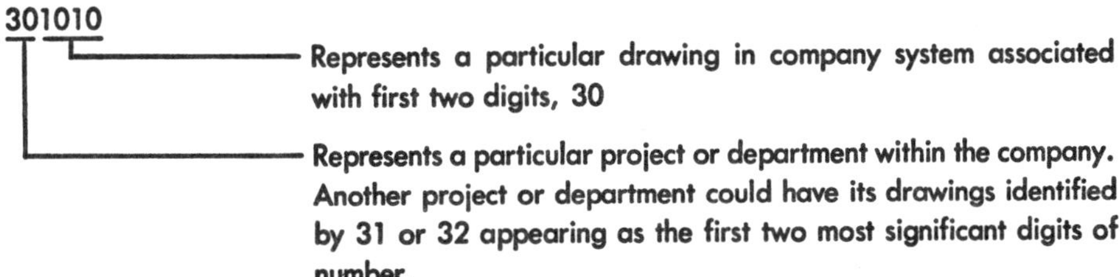

DISADVANTAGES AND ADVANTAGES OF TWO SYSTEMS

The significant and nonsignificant systems have their adherents and detractors in industry. While the coded system seems to be a useful approach, its success depends on a few critical factors that are hard to control. For example, if a complex coding system is to be useful, it requires the ability to be decoded quickly. That is, those involved must know the codes so well that they can quickly determine information about the part or document without spending much time finding a code definition list or asking the originating department for help. This condition is often hard to attain because people (1) won't spend the time needed to memorize the codes; (2) misplace their coding sheets; and (3) rely on their memories, which are often fallible, resulting in errors. Also changes in codes such as additions, deletions, and corrections may continue for a long time after the initial definition of the system. Thus the staff is required to be continuously learning new codes.

Another problem is that it takes time to properly identify a part or assembly with the code number when the drawing is being prepared. In contrast, the draftsman need only go to a log book for the next number available when a nonsignificant numbering system is used. In addition, a complex coding system has more chances for errors and requires a stable staff to provide the desired benefits.

Another disadvantage of significant numbering systems is that they offer fewer useful numbers than an equivalent length nonsignificant system. For example, if a completely nonsignificant numbering system is used, each number is issued in sequence as a drawing is prepared and there are no wasted numbers. However, when a significant system is used, there are always some numbers within each classification that are never used. The following illustration demonstrates a 40 percent loss of drawing numbers on one project:

System: uses first 3 digits to identify project or model number of product. Next 4 digits are used to sequentially assign numbers to drawings as they are prepared for project.
Sample project drawing block assignment: 4010000 to 4019999
Total available drawing numbers: 10,000 (first 3 digits are the same for all project drawings: 401)
Drawings required for project: 6,000
Drawing numbers wasted: 10,000 − 6,000 = 4,000

Why is this bad when a 10 or 15 digit number system can be defined? It is to the company's advantage to keep the lengths of its identification numbers as short as possible because recording and working with large numbers is more expensive. Also more errors can be made when large numbers are used. Of course, a short number system can become saturated, requiring a new system in 10 or 20 years. This is also undesirable from a cost and continuity viewpoint. Therefore, it is important to avoid wasting drawing numbers once a fixed set of digits for each identification number is defined.

Because of the preceding disadvantages, the cost of developing, implementing, and controlling a significant system can be high and the anticipated benefits may not be realized.

The advantages of a significant numbering system, however, may cancel its disadvantages. The ease of translating the number of a part or drawing into an approximate mental image of its configuration and its relation to higher and lower subassemblies can save a user time under some conditions. For example, a stock clerk may immediately know what a part looks like and its storage location by simply reading the part number on a material requisition form. In addition, a designer or engineer may want to find a standard part for a particular application. By knowing the coding scheme, he can refer to the general class of part desired. (Of course, a detailed catalog index can be used for this purpose also.)

The disadvantage of a nonsignificant numbering system is that it tells nothing about the configuration, level of assembly, or purpose of the item depicted on the drawing. However, the system is simple to create, implement, and maintain. It also has a potentially longer life span.

Some coded information can be automatically obtained by assigning blocks of numbers to different projects in sequence. Thus, drawings or parts can be identified in this way. Once the number block is issued, the draftsman simply assigns each succeeding number to a drawing as it is prepared.

CRITERIA FOR SELECTING A NUMBERING SYSTEM

Before describing the criteria for selecting a numbering system, note that if a company already has a system in effect, it should not be tampered with unless it is unworkable as is. It is better to live with a poor but workable system, than to change over to a new system. The confusion, extra work, and errors that will result by trying to change a numbering system that is already in operation will have a serious impact on company operations. Of course, if a new company is being formed or a physically separate and independent division is being created, a new system can be developed. (Divisions within the same physical location should use the same system.)

In the design of a numbering system, certain practical criteria must dominate its configuration. These include ability to satisfy all documentation and part numbering needs on a company-wide basis, simplicity, brevity, convenience in use, uniformity, versatility, distinctiveness, and maximum length. These qualities are discussed next.

SATISFYING COMPANY NEEDS—The system must be designed with regard for all documentation and parts identification needs. It should also have sufficient capacity to allow unique identifiers to be issued for many years to come. The system must be satisfactory to D&D, data control, manufacturing, data processing, material control, quality assurance, and engineering. Consequently, it should be coordinated with all these groups before final preparation and release. Therefore, D&D personnel should not develop a system on their own simply because no one in higher management has implemented a company-level numbering system. (The D&D manager should contact other departments and take the lead in the absence of standards or configuration management groups and formulate a numbering system.)

After the inputs from each department, if any, have been obtained, the system can be designed with the following criteria in mind.

SIMPLICITY AND BREVITY—The simplest and briefest system is best if all other constraints or needs are met. A simple and short numbering system offers the following advantages:

- Easier to design, understand, implement, and maintain
- Easier to remember system
- Minimizes errors in recording, alterations, and recall
- Increases productivity because a short, simple identifier takes less time to write down, to remember, and to mark on equipment. It also needs less space to record on documents and fewer memory spaces in the computer.

CONVENIENT TO USE—A convenient system is one that is easy to use. Simplicity and brevity, of course, make a number system more convenient to use. But other factors apply such as ease of obtaining new numbers, applying them to documents, and retrieving documents with them. Thus, while a simple and uniquely structured numbering system is inherently easier to use than a more complex and random system, the overall procedure for creating and using numbers to identify documents and parts must also be easy to do.

UNIFORMITY—Numbers should have a common format throughout the company, especially numbers used for drawings and associated lists. Different departments or divisions should use the same formats to assure consistency of interpretation and use. Thus, one department should not be permitted to identify its drawings with a 7 digit number while another department uses a 9 digit combination of numbers and letters.

DISTINCTIVENESS—Different formats should be used for different types of documents. For example, drawings, specifications, reports, and parts lists should be easily and positively distinguishable from each other. Identifiers can be made distinctive by combining numbers and letters, using acronyms, breaking up identifiers with dashes, or by using slashes. For illustration, military specifications are readily recognized by their distinctive structure: MIL-C-43322/4. Military standards, on the other hand, use a different center element and shorter numbers. Thus: MIL-STD-100.

VERSATILITY—The basic numbering system should be able to take care of new requirements. For example, to identify a new type of document or product not originally planned for. Thus, a new type of specification should be easily identified with a unique number without revamping the system.

INTEGRATED—Sometimes it is convenient and efficient to develop an integrated number system that ties various related documents together with a common number. An example of an integrated system that connects drawing, parts list, artwork, and test specification for a single item together is shown next:

Document Type	Identifier (all identifiers contain 301000)
Top assembly drawing	301000
Parts list	PL-301000
Master pattern/artwork	MP-301000
Printed circuit/wiring board assembly	PC-301000
Wire list	WL-301000

Design specification	DS-301000
Design layout	DL-301000
Test specification	TS-301000
Acceptance test procedure	ATP-301000
Final Report	FR-301000

Construction drawings frequently follow the same integrated system using bills of materials, valve list, and other construction calculations/documents.

MAXIMUM LENGTH—A maximum length of 15 symbols is set by the Government for drawing and part numbers. This restriction is applied to allow all numbers, regardless of their source, to be used in Government data processing systems. That is, a fixed field of 15 spaces is allotted for data processing cards. If a number exceeds this amount, there is no room for it. Therefore, because most companies do some work for the Government and because a number longer than 15 symbols is cumbersome to work with, avoid exceeding this limit. Of course, internal company constraints on identifier lengths may also apply. Therefore, data processing and other departments may restrict maximum lengths of identification numbers.

CASE IN POINT: In addition to practical and Government length restrictions, handling large numbers costs money. For example, the recording of a 10-digit number will require about 50% more labor and money than a 5-digit number. Even for a small company, a long number system can increase costs by more than $1000 per year. These numbers are recorded on document status cards, engineering orders, purchase requests, indentured drawing lists, and as-built lists. Thus many thousands of numbers are recorded every year.

For medium and large companies, the costs can exceed several thousand dollars annually due to increased recording labor, computer time, errors, and marking of hardware.

5.2 Drawing, Part, and Document Numbers

Drawings must be identified by a unique number* to obtain positive, unambiguous, and permanent identification of each document prepared. No drawing prepared by the D&D department should ever be released without an approved and controlled number recorded on it.

After a number has been assigned, it should never be reassigned to another

*The term "number" includes alphanumeric identifiers.

drawing even if the original drawing is destroyed. There is always the possibility that a stray copy of the old drawing may be taken for the newer document carrying the old number. In addition, the cost of assigning a new number is trivial when compared to the labor costs in building and testing the wrong part.

Because part numbers are often the same as the drawing numbers or contain the drawing numbers as an integral part of them, both drawing and part numbers are considered here. The following sections describe some drawing and part number systems. Note that these systems are used for illustration and that a wide variety of systems exist in industry.

BASIC DRAWING NUMBERS

A basic drawing number usually contains from 4 to 15 alphanumeric elements, depending on company needs and philosophy. Examples of three numbering systems follow:

Example	Description
123456	Pure nonsignificant number
ARC-10000	Alphanumeric; partial coding (ARC stands for company or system and 5-digit number is issued sequentially)
FD-B-123456	Alphanumeric; partial coding

As can be seen, the length and format of numbering systems can vary considerably and may include the drawing size and year of preparation if desired by the company.

When the drawing is changed, a revision letter is placed after the basic number. For example, the first revision to a drawing is labeled A, the second B, and the third C. The revision letter looks as follows: 123456A. Note that while the revision letter is an essential part of a drawing, it is not an integral element of the part number and is therefore not used to identify the part. However, many drawing number systems use letters for purposes other than to indicate revision status.

When extensive revisions are made to a drawing, and all letters are used up in the alphabet, a second row of letters may be used: AA, AB, AC, AD, etc. Certain letters are forbidden for use as revision letters because of their tendency to be confused with numbers or other letters. These include:

I (could be taken for 1)
O (could be taken for zero)
Q (could be taken for "O" or zero)
X (could be taken for a handwritten Z with a bar across it)

Although S and Z are usually not included in the forbidden letter list, an S can be confused with a 5 and a Z with a 2 or a 7 when handwritten. Thus, as long as typed letters are used no problems will occur.

BASIC PART NUMBERS

Part numbers are important because they control the assembly and replacement of items within end items. Therefore, their issuance and control are essential operations.

Part numbers can range over a wide number of digits and letters. Some companies have used part numbers with 17 symbols. However, as mentioned, part numbers should be kept at 15 symbols or less so that they can meet government data processing requirements. The drawing number is usually the same as the part number included within it. This is desirable so that the documentation for a part can be easily recovered and referenced to it. A few types of nonsignificant part numbers are presented next. Again, the formats shown represent only a few of many varieties in use:

Example	Description
123456-1	Part number, one suffix
123456-2	Part number, new suffix
123456-1-2	Part number, two suffixes
123456-101	Part number, one 3-digit suffix

The first number is marked on a part to identify it. When the part is modified so that its functional and mechanical characteristics make the old part noninterchangeable with the new one, a dash number is stepped to indicate this condition. If the part is altered again, a new dash number, increased in magnitude by 1, is used on the part. If a dash number is not used, the part would have to be assigned a new number such as 123457 if this number is unassigned.

A dash number is often used to indicate one of several parts on a tabulated or multidetail drawing (see 4.1). A single number suffix is used for this purpose just as shown for the second example given. When this condition occurs and the part is changed so that it is no longer completely interchangeable with the first part, a second suffix can be added as shown in the third example. In this case, the dash represents the following:

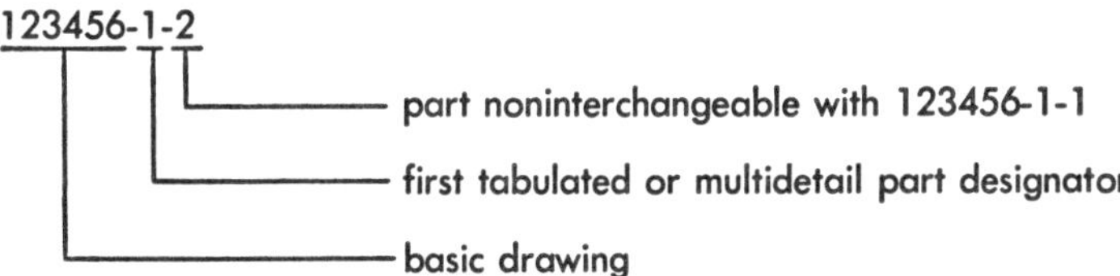

Other techniques would be simply to step -1 to -2 or to the next available number without using a second dash number. Thus: 123456-2 or 123457-3.

Suffixes need not be limited to single digits. They are often 3 or 4 digit numbers and coded to indicate the indenture level of the item. A possible format for this type of part number system follows:

50000-500-XX Top assembly (XX can be 00 to 99 for all assembly levels, including those listed below)
50000-400-XX Major assembly below top assembly (-400 to -499)
50000-300-XX Subassembly below major assembly (-300 to -399)
50000-200-XX Component below subassembly (-200 to -299)
50000-100-XX Piece part (-100 to -199)

The numbers given above represent different tiers in a 5-level indenture system. For items in the same level, the next larger 3-digit number is issued. Thus, four major assemblies in the second highest tier would be numbered: 50000-400, 500000-401, 50000-402, and 50000-403. Note that the system can be reversed so that the highest assembly level starts with a -100 and the lowest with -500. Table IX gives some other examples of part numbering systems.

Table IX

Examples of Part Numbering Systems

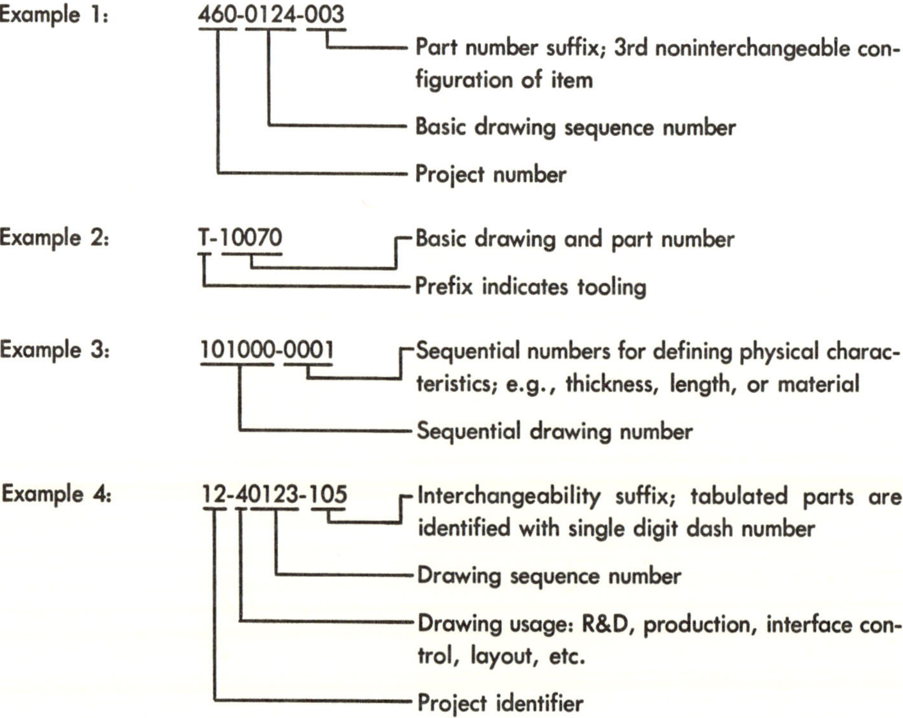

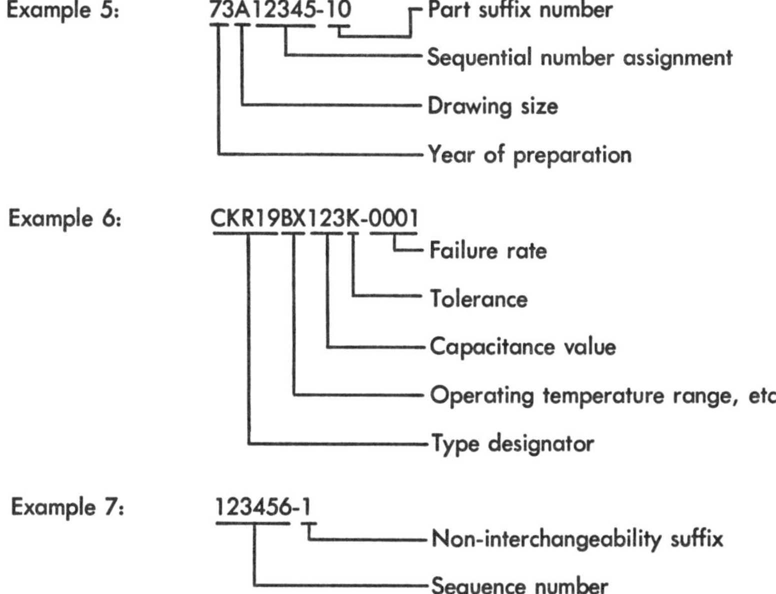

As mentioned, revision letters recorded on the drawing are never recorded with the part number. Revision letters represent document changes, not hardware changes. Also, a revision letter may merely indicate that a note was added to the drawing or that a spelling error was corrected. Thus, when a change makes a part noninterchangeable, a number suffix or entire part number change is made.

Note that a letter change in the end item or model number identifier of a product does not conform to this rule. The letter added after a model number is not a revision letter and is therefore not subject to the above restriction. These numbers are discussed later in this chapter.

A SAMPLE CODED DRAWING/PART IDENTIFICATION SYSTEM

When it is desired to identify various levels of assemblies and drawings for a complex product, a coded numbering system can be used. For example, the product can be identified as illustrated in Figure 31.

FAMILY DESIGNATION NUMBERS

Family designation numbers are permanent identifiers used when items have similar configurations and are used for the same application. They provide a stable base for identifying items even through their configurations may be modified so that they are no longer interchangeable. Serial numbers are used

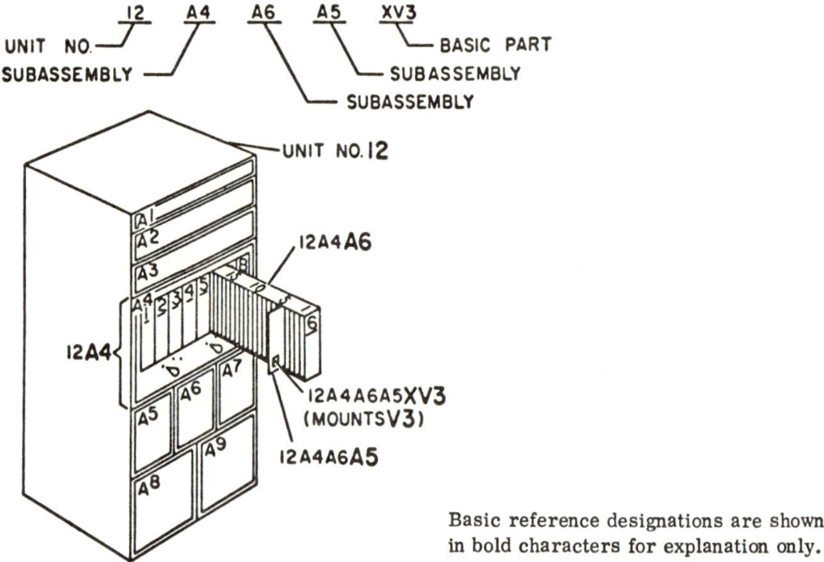

Figure 31 Identification of Various Assembly Levels in Equipment

to identify each item in the family of units built during a project. An example of a family designation number (FDN) is:

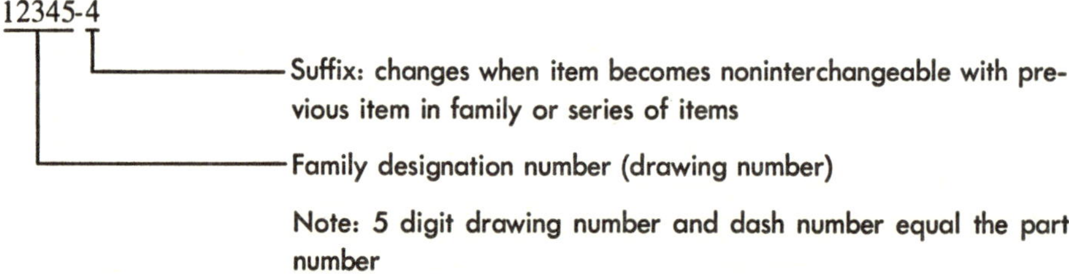

One advantage of the FDN is that it enables identification of noninterchangeable items in the same family. Thus it enables easier determination of the number of items built for a product and how many spares for each noninterchangeable item are needed.

Another advantage is that it provides a permanent base for serializing items to meet quality assurance and supply requirements. All items are sequentially identified by serial numbers with respect to the FDN even through they are reidentified with a dash number to indicate noninterchangeability. The combination of FDN and serial number simplify identification of items intended for the same use and allow systematic evaluation and rework of items requiring a changed configuration.

DRAWING TREE FOR ILLUSTRATING PRODUCT BREAKDOWN

Family drawing trees are useful devices for identifying, interrelating, and assigning numbers to all the drawings used in a complex system. They are similar to organizational block diagrams and have information flows from the top to the bottom of the sheet. Blocks are used to indicate the drawing title and number. Each successive block leads to a row of drawings that represent a lower assembly level as shown in Figure 32.

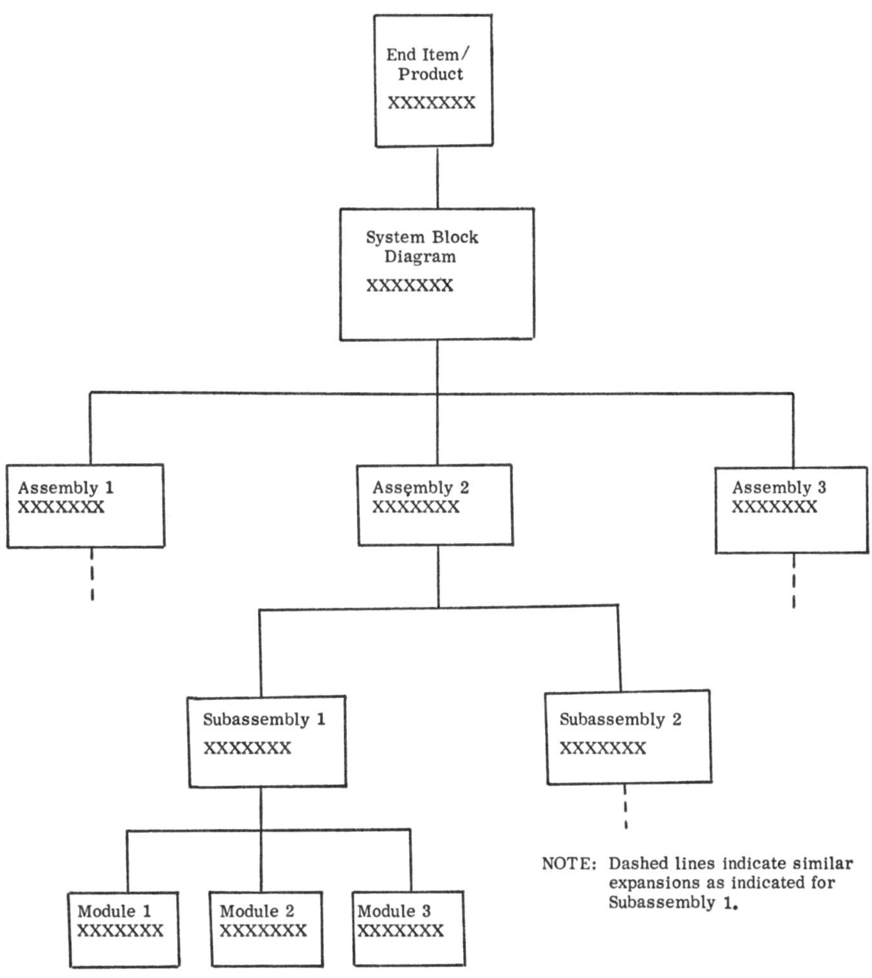

Figure 32 Drawing Family Tree

The family drawing tree for a system or product is prepared when the basic concept, block diagram, and overall system layout are completed. Equal level assemblies or subassemblies are shown on the same horizontal line.

While drawing trees provide an excellent graphical presentation for identifying the drawings required for a product, they are not always used. Sometimes an indentured drawing list is prepared from the system layout drawing. (See Section 5.1.)

5.3 Assignment of Numbers

Assignment of numbers to project or products within one company should be under the centralized control of one department or group, preferably the data control facility. This method allows absolute control of the numbers issued and their use.

CASE IN POINT: One company had two major divisions located in one facility, each with its own number issuing system. As a result, duplication of effort was a frequent occurrence. For example, similar parts were assigned different standard control numbers and control sheets for assuring interchangeability of parts received from the vendor. (Of course, this occurred without each division's awareness of what the other was doing.)

The impact on the company was that the same parts were identified differently, documented separately, ordered separately, and stocked in different bins. The result was lost money due to smaller quantity discounts, more purchasing labor, more stocking and material control effort, and artificial shortages. (If one bin was empty, a new order was made even through identical parts were available stored in a nearby but differently numbered bin.)

To obtain a block of numbers, the D&D manager or chief draftsman completes a form defining his need, project number, and authorization (approved job order). Upon submittal of this form, the data control clerk assigns a block of numbers sufficient for his anticipated needs. The block of numbers issued is recorded on the form and in a log book that identifies all drawing numbers issued since the company began operations. It also contains unassigned numbers listed in sequence. Thus, issuance of a block of numbers is made from the available, unassigned numbers.

Once the D&D department obtains its numbers, the designer/draftsman responsible for the project controls the assignment of individual drawing numbers. He records the drawing title, name of draftsman issued number, and date of issue next to each number in a project drawing log so that he knows exactly which numbers have been issued.

The disadvantage of the above system is that it can be wasteful of numbers. To ensure that enough numbers are available, more numbers than needed will be requested. Even though the unused numbers may be returned to data control for reassignment, another drafting supervisor may prefer a new block of numbers for his project. To avoid this situation, a draftsman can call data control each time he needs a drawing number. The data control clerk then gives him the next available number in the log book and records the title and other information next to the number. The disadvantage of this system is that drawings for a project will not have a common base. For example, if 1000 drawing numbers are issued for a project/product, most numbers will be completely unrelated and mixed with numbers assigned to other job orders.

5.4 Other Types of Configuration Identifiers

Besides drawing and part numbers, different types of identifiers are used to control the configuration and related documentation of an item. These include:

- Specification numbers
- Serial numbers
- Configuration item (CI) or end item numbers
- Change identification numbers
- Code identification numbers

The purpose and characteristics of these identifiers are described in the following sections.

SPECIFICATION NUMBERS

Specification numbers are key identifiers that must be carefully controlled and recorded. The format and length of these identifiers depend on the numbering system used. If an integrated numbering system is used, the specification number may be the same as the top assembly number of the end item with a prefix letter used to designate it as a specification number. Of course, if the specification has been prepared before the top assembly drawing has been prepared, the drawing number won't be available and another number will have to be used. Some number format examples were given in Section 5.2. A few more are listed below to illustrate the wide variety of specification number formats:

S-123456B
ST-10500 (represents a standard part specification)
S-530-P-0028
ZPE-1081-0002A
50M60154

Specification numbers are modified by adding letter revisions to them each time the contents are changed. Unless radical changes are made to the document, the original specification number is retained.

SERIAL NUMBERS

Serial numbers are used to distinguish among various items of otherwise nearly identical parts within the same family. Thus, if 10 items with part number 123456 are built, each one has a sequence number (serial number) added to it to provide a separate identity for it. Thus the first part built is identified by 123456 SN 01, the second 123456 SN 02, the third by 123456 SN 03, and so on. Some serial numbers have letter prefixes to differentiate among different usages of a product; e.g., SN BFE 1234 for production units, SN X001 for experimental units and SN P001 for prototyped models.

The advantages of serial numbers are:

1. Allow engineering/manufacturing to determine point at which changes will be incorporated in the series of items built. (Also allow tracing changes back to specific items after they were incorporated.)
2. Allow cross referencing manufacturing, test, and inspection data to a particular item in the family.
3. Allow more effective accounting (e.g., in stock, shipped, returned to supplier, or modified) of various items.
4. Allow control of installation and location of specific items within the family. (Although parts with same number are supposedly the same, design/manufacturing changes may have been made that require further modification of particular items in family. Without serial numbers, it would be nearly impossible to identify the items that need to be changed or have had variations made to them but are still interchangeable.

CASE IN POINT: One company failed to define its serial number system in writing. As a result a firm standard practice did not exist and various attempts to implement a system resulted in numerous false tries and meetings with high level company managers from purchasing, material control, production, data processing, quality assurance, and engineering.

After several years and the expenditure of over $5000 in lost labor, an approved procedure was still not implemented and the company could not tell what was the last serial number issued against a number of assemblies used in its equipment. That is, no one knew what the next serial number should be for assignment to the next part or assembly that came off the assembly line. Thus, either duplicate serial numbers would be marked on an item (two items would have the same serial number) or a break in serial numbers would occur on the next order received from a customer. As mentioned before, skipped serial numbers is not an acceptable practice in industry.

CHANGE INDENTIFICATION NUMBERS

Change identification numbers are assigned to documents that describe alteration to the configuration of an item on a drawing or associated document already released for production. These change identification numbers can take many forms, depending on the type of change document. Different numbering systems are often used for waivers, deviations, engineering change proposals, engineering orders, advance design changes, and drawing change notices. Some examples follow:

Document Type	Number Format
Waiver	W 01
Engineering change proposal	ECP-15, -16, -17, etc.
Engineering order	EO 1060
Advance design change notice	Drawing number plus a slash number: 123456/1, 123456/2, etc.
Drawing document change notice	DCN 123456A or DCN 1234
Specification change notice	SCN 10-17B
Procedure change notice	PCN 5-9a

PCN 5-9a:
- 9 — Sheet 9 of procedure; a is first sheet changed, b the 2nd
- 5 — Fifth change to procedure

CODE IDENTIFICATION NUMBER

Code identification numbers are government issued identifiers for companies that control design and manufacture of equipment on government contracts. A number assignment can be obtained if requested by a company doing business with the government. Write to:

Defense Logistics Services Center
Attn: DLSC-CGC
Federal Center
Battle Creek, Michigan 49016

The code identification number is a 5-digit identifier uniquely related to the company and recorded in *Military Handbook* H4-1, "Federal Supply Code for Manufacturers, Name to Code." Large companies with several divisions have more than one code assigned to simplify identification of the equipment source. The code identification number is recorded in the title block of all company drawings and associated lists. Code identification numbers are also recorded in parts lists next to each item that is not produced by the company (when required by contract).

5.5 Identifier Characteristics and Limitations

Identifiers can be pure numbers or combinations of numbers, letters, and other symbols. They rarely, if ever, are pure letter combinations. Some restrictions are generally applied to identifiers for increased reliability, accuracy, and ease of use. These restrictions, some of which have already been described, are summarized below:

- Letters I, O, Q, and X are not used because they can be mistaken for numbers, similar letters, or for different handwritten letters or numbers. (See Section 5.2.)
- All letters should be capitals, including revision letters.
- Numbers should be whole arabic numerals (no fractions, Roman numerals, or decimals are allowed).
- Blank spaces are not used. Separation between parts of an identifier is made with a dash; e.g., A4-4000. (Note that this practice is not always followed.)
- Symbols such as #, /, *, etc. are not used for drawings and part numbers; however, slashes are often used with military specifications and advance design change notices.
- Total number of symbols should be less than 15, including dashes.

REVISION LETTERS

Revision letters are essential elements of identifiers for documents. Whenever a document is altered, the identifier should have a new revision letter added. Note that a spelling change as well as design change both merit a revision letter change although the first one seems trivial. The reason for this practice is to avoid errors in judgment when gray areas are entered. For instance, a simple spelling correction won't affect the configuration. Therefore, it is safe to avoid adding a revision letter to a drawing. However, what happens when an innocuous appearing note is added by the draftsman that actually represents a significant change in processing or manufacturing requirements? Failure to add a revision letter to flag the change can result in a serious oversight by the manufacturing engineer because he has already reviewed the drawing and the new drawing has no change in identifier to indicate that a change has been made. However, when this change control is a well-defined responsibility, revision letters may be omitted for minor drawing alterations.

5.6 Physical Identification of Items

The physical identification of parts and assemblies is not the responsibility of the D&D department. However, instructions for marking hardware correctly should be recorded on each drawing to ensure that this important step is not omitted during fabrication of the item. If the method of marking must be made to a particular specification, the complete specification number and title should be given on the drawing, usually under "Notes."

When a specification is not used to mark the part number on the item, the lettering style, size, and method of implementation must be given. If a protective coating over the number is needed, this should also be specified. Of course, the location of the part number is important. It should be placed on the item so that it can be seen after assembly into its next higher level assembly.

The method of marking depends on the usage of the item and its environment. Methods for marking items are listed below: (Tags are used when parts are too small to mark.)

1. Rubber stamp
2. Steel (impression) stamp
3. Engrave
4. Decal
5. Paint or ink
6. Identification plate

Part and assembly numbers should also be identified on the drawing when the part number is different from the drawing number. This number should be recorded above the title block. When a separate parts list is used, the part number, including dash numbers, is recorded in space allotted for this entry or is recorded as the first part entry in the list

5.7 Reidentification of Parts and Assemblies

The design engineer is primarily responsible for indicating that a change to an item requires part number reidentification. However, the designer and draftsman are key men for ensuring that this is done when necessary. Failure to reidentify a changed part can lead to expensive and time consuming problems when the product is delivered or put into operation. For example, the field engineer may replace a defective assembly with a redesigned assembly that has the same part number but is not interchangeable. As a result the product may be damaged or will not work right, causing additional troubleshooting, delays, and customer dissatisfaction.

Needless reidentification of an item also leads to extra documentation costs, surplus stocking of the same items with different numbers, higher inventory taxes, and artificial shortages. Therefore, D&D personnel should always be alert to errors in parts identification.

An item should be reidentified only when its functional or physical characteristics are changed, preventing it from being completely interchangeable with its predecessors. When a part is determined to be noninterchangeable, either a new number should be issued or its suffix stepped up to the next larger value. In general, however, if its predecessors can be reworked to the new configuration, only a suffix change is needed. (Note that the capability of being reworked to the new configuration does not mean that the older items must be reworked before a suffix is added—only that they have the potential for being reworked.)

The importance of reidentification is illustrated below.

CASE IN POINT: In one company, failure to reidentify cables that were found to be unsafe resulted in the inadvertent delivery of previously built cables with the same number. The use of these cables by the customer's personnel caused one of his people to receive a near lethal electrical shock.

Although the company's engineers thought that they had corrected the safety defect in all the cables built, they missed several cables that were in a different location for some additional processing. When they were returned to the stock room, they were added to the already reworked cables. However, since the reworked cables had the same part numbers as the defective ones, there was no way to distinguish between them.

If the company had issued a new number, or at least a dash number, and corrected its documentation so that this

new part number was called out, this accident would have been avoided.

Rules for guiding D&D personnel in part number reidentification follow. Change the part number when these conditions exist:

1. Items are altered so that old and new parts are not directly and completely interchangeable in regard to installation, safety, or performance.
2. Old items are limited in use to specific serial numbered products, but new items can be used in any product or end item.
3. Performance or durability is affected to such an extent that old items must be scrapped for safety, poor or faulty operation, or low reliability.
4. The company alters or selects an item that has been identified and documented by another company.
5. A material, process, or protective treatment is changed so that any of the conditions described above exist.
6. An item is reworked, using a modification kit, into a later dash number version of the item and is completely interchangeable with all items identified by the later dash number.
7. An item is established by the company as a standard and is identified by a standard specification identification number when the following conditions apply:
   a. The item has a multiple usage and is expected to be used in more than one product or end item.
   b. The item is not repairable and spares will not be kept by the customer below the level identified by the standard specification identification number.
   c. The item is completely defined in a specification, source control drawing, or specification control drawing, including performance, durability, reliability, form, fit, quality, and inspection requirements.
   d. More than one source is approved and qualified to supply the item.

Items covered by approved standards and used without alteration or selection should be identified by their standard part numbers.

Note that when a repairable assembly contains a noninterchangeable item, the item and its parent assembly are reidentified with new part numbers. Also, all higher assemblies up to and including the assembly where interchangeability is reestablished are reidentified. The interchangeable higher assembly is reidentified because its parts list contains noninterchangeable items and thus has a different internal configuration from the point of view of repairability.

Sometimes a part may be functionally and physically interchangeable with its predecessors but may still require reidentification. For example, a panel meter may have had its face altered in shape and color but all other factors have remained the same. Since users may have already designed their products using the old configuration, this new part should be renumbered so that they are not sent the different appearing meter, unless they know what they're getting.

Some parts require reidentification with completely new numbers. Suffixes are not added to the original number to indicate noninterchangeability. New numbers should be issued for the following items:

- Altered parts: an existing part's configuration is changed
- Selected parts: parts are selected for special characteristics
- Matched set: two or more parts are mated together and are permanently associated with each other

6.0 Control of Data and Changes for Maintaining Integrity of Product Configuration

Hundreds of dollars are often invested in the creation of an engineering document. However, this investment can become worthless without controlling the integrity of the information recorded on a drawing, parts list, or procedure. A drawing involves much thinking, calculating, refining, and experimenting before it can be used for producing an item. Therefore, once released to manufacturing or purchasing, the information on the document must be protected from uncontrolled changes. This process of controlling changes has become increasingly important in modern industry because of three prime factors: (1) complexity of the products being created, (2) interrelatedness of different products and items within products, and (3) rapid pace of technological and design changes. As a result, change control has become a necessary discipline that involves following well-defined procedures and considerable effort on the part of managers, engineers, designers, checkers, and draftsmen.

CASE IN POINT: One medium size company on the West Coast spends about 30 percent of its design and drafting labor on changes to engineering data. This is not unusual for many companies involved in development of highly complex equipment and may even be greater for certain projects and companies.

With this much work being done in relation to design and data changes, everyone must be prepared to deal with changes efficiently to minimize delays, defective products, slipped schedules, and higher costs.

BENEFITS AND APPLICABILITY OF THIS CHAPTER

The benefits of this chapter include a survey of change procedures and activities needed to ensure control of product configuration. Other benefits include:

- Systems viewpoint of change control cycle.
- Descriptions of change control documents used in industry.
- Procedure for revising engineering data.

Implementation of formal procedures for controlling changes can avoid unnecessary costs trying to determine what the original configuration was before it was changed.

> **CASE IN POINT: One large company spent over $70,000 trying to reconstruct the configuration of a complex equipment because its engineers failed to document changes made during the final evaluation and checkout phase of the effort. The solution to this type of problem is an explicit change control system and strong discipline to make it work.**

Because changes to engineering data occur in all organizations, this chapter applies to all size companies.

6.1 Change Control Cycle: A Systems Viewpoint

The change control cycle involves many steps and people. However, the exact nature of this cycle depends on company and customer requirements. Figure 33 is a typical change control flow diagram.

The basic change control process consists of six major functions: (1) determination of the need for a change, (2) identification and logging of change, (3) description and documentation of change, (4) evaluation and approval of change, (5) incorporation of change in hardware, and (6) verification and documentation of change incorporation. These operations are described next in terms of the change control cycle shown in Figure 33.

ORIGINATOR—determines need for a change due to design error or improvement, unavailable parts, equipment malfunctions, or cost savings. Marks up document to show desired configuration, change cut-in points, and disposition of existing parts. Obtains signature of responsible engineer on change document. Submits change package in special folder (identifies it as a formal change) to Document Control Group.

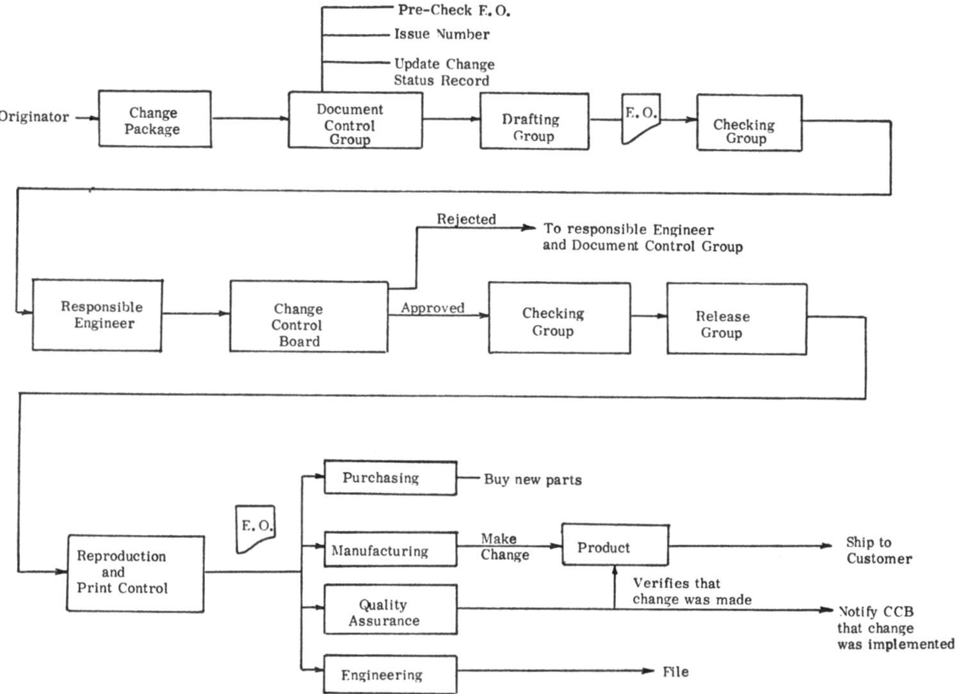

Figure 33 Change Control Cycle

DOCUMENT CONTROL GROUP: inspects package for completeness and legibility. Also verifies that other documents affected by change have change packages submitted. Assigns unique change document control number and logs in master change number log book. Also updates change status record, which identifies all changes in process or released against each document. This group may reidentify an EO as a CN if a direct change to vellum (original) is needed because of change complexity or outstanding EO's exceed five. (See 6.2.) Note that if any EO/CN's are already in process, these are checked against the new EO/CN for conflicts.

DRAFTING: prepares change document (e.g., engineering order or change notice) to engineer's marked-up print. Both pictorial and textual changes must be recorded on change document. Both the old and new configurations are usually shown on the change document. Any conflicts or ambiguities are brought to the designer's or engineer's attention.

CHECKING: Verifies that change document is prepared to standard company practice. Ensures that drafting has prepared change document as required by marked-up print from engineering. Makes sure that all information needed for change document is recorded correctly. Compares interrelated change documents (e.g., assembly drawings, detail part drawings, parts lists, and layouts) to assure consistency in description of change and in design configuration. Contacts engineer or designer when technical discrepancies are discovered.

RESPONSIBLE ENGINEER: verifies that the change was made as required. Corrects any errors he may have made and those made by draftsman. Approves EO/CN's for release.

CHANGE CONTROL BOARD: a change control board (CCB) is used by many companies to evaluate all changes made to a document already released to production or purchasing. The use of a CCB depends on the size of the company and the need for coordinating changes among various interdependent groups; e.g., a product change may affect production, quality assurance, purchasing, publications, testing, and maintainability groups. Therefore, these groups should review and comment on the proposed change to ensure meeting schedule, cost, and customer requirements.

The CCB reviews each change for the following:

- Impact of change on overall effort and specific departments
- Interchangeability and reidentification of revised part with other similar parts
- Feasibility of making change and scheduling problems
- Value of change in terms of performance and cost
- Effectivity of change; i.e., at which serial numbered assemblies/end items should the change be cut in.
- Disposition of existing, in-process, or in-use items

After evaluating and approving the change, the change document is signed and returned to the Release group. If the CCB finds the change unacceptable, the document is revised or returned to the responsible engineer for correction or cancellation. The status of the change is reported to the Document Control group.

RELEASE GROUP: verifies that all needed information is on the change document and the required signatures have been recorded. Adds change letter to document and records this information on change status record. Submits change document to Reproduction and Print Control for issuance to staff. (In some companies, status change data is input to a computer which provides summary reports on the revision and release status of all project or company engineering documents.)

REPRODUCTION AND PRINT CONTROL: updates the document status record to reflect latest change against each document. It also prints the required copies, stamps them "Production Release" and issues them to staff. Files original change document for future reference.

PURCHASING: responds to new material requirements if specified on change document. Notifies engineer or CCB of problems not previously identified.

MANUFACTURING: implements change to product and records data on travelers. Records actual cut-in points for changes in hardware.

QUALITY ASSURANCE: verifies that change was made to product and meets requirements of change notice. Reports results to CCB when required.

ENGINEERING: reviews change document when new changes are proposed or needed to avoid duplication of efforts or incompatible alterations. Assists manufacturing when problems occur during implementation of changes.

The elapsed time for implementing the preceding change cycle for a typical document can range from a few hours to a few months, depending on the size of the company and its mode of operation. Because of this delay, there is often much pressure to circumvent the system. However, experience indicates that short cuts usually cause more problems than they solve and often don't save time in the long run. Of course, an emergency change cycle may be needed for special documents or conditions. A short cut system is acceptable if it is well-defined and controlled. However, because of the risks and added costs that are usually incurred from an emergency system, it should be used only when delays will prevent critical product deliveries, cause the company to lose money, or jeopardize the company's position in the award of a new contract.

When the complexity of the change control cycle seems unacceptable because of long delays in issuing change documents, verify that each person in the approval loop is really needed to assure proper evaluation of the change, its accuracy, and its documentation. If someone is merely rubber stamping changes, eliminate him from the cycle or find someone who can make a meaningful contribution to the change document.

6.2 Document and Change Control Documents

Document and change control documents are essential for controlling the integrity of engineering data and changes made to them. Some samples of these documents are discussed next.

INITIAL RELEASE DOCUMENT

A special form is needed whenever an engineering document subject to formal control is released by engineering for use by production or purchasing. This document is called a release and issue form. It is filled out by the Release

ENGINEERING RELEASE AUTHORIZATION

SH OF

NUMBER EA GO —

TYPE OF RELEASE: ☐ ADVANCE ☐ PRODUCTION

DATA RELEASE IS: ☐ PARTIAL ☐ COMPLETE

EQUIPMENT NOMENCLATURE

PREP	DATE	CHECK	DATE	ENGR	DATE	ENGRG PRGM MANAGER	DATE	REL ISSUE DATE

CONTRACT CODE NO

TC	PRE	DRAWING/DOCUMENT NO	REV	SIZE	SH	NOMENCLATURE	NEXT ASSY
RA							

DISTRIBUTION/REMARKS

Figure 34 Release Form

Group when the drawing, or other document, has been checked and signed by the required people.

Figure 34 shows a sample Engineering Release and Issue form. Several documents may be released by it. Its purpose is to provide administrative and control data for the Print Control and Reproduction Group and a permanent and official record of engineering authorization for manufacturing to use the document for production. (Note that Figure 34 shows a form that is used to input information into a computer.)

The release form and associated document or documents are submitted together to Reproduction. This group verifies that the Release form is signed by the correct people and then fills out a document history and status card. Copies are then made and distributed to the people designated on the Release form.

Note that the Release form contains information on physical characteristics of the document (size, number of sheets, and next assembly) and identifies work order (GO/EA) and contract. Other information includes whether it is an advance or production release. (These categories were discussed in Chapter 3.)

The extent of the information recorded on the Release form depends on company needs, policy, and customer requirements.

DOCUMENT HISTORY AND STATUS CARD: KEY TO DATA TRANSACTIONS

The document history and status card, as described before, is prepared upon initial receipt of a released drawing. Its purpose is to create a single source of information on each document released to manufacturing, test, and purchasing. Data items recorded on this form include the date, document number and title, originator's name, product or project number, next higher assembly (NHA) number, size, and number of sheets.

Whenever a change is released against this document, it is updated to identify the date of revision, the change document number, distribution code, and associated change letter.

This is a critical document and is the basis for informing engineering, manufacturing, inspection, and purchasing of the latest revision existing against that document.

These cards are maintained by Reproduction and Print Control (also called Data Control) personnel and are not allowed to leave the area or to be touched by anyone but an authorized data clerk so that the cards won't be incorrectly changed or lost.

CHANGE REQUEST FOR FOCUSING ON TROUBLESPOTS

After initial release of a document, changes may be initiated by submitting them on a document change request form as shown in Figure 35. The purpose

Change Request	Date	CR ______
	Revision	

Contract Number ______ CI P/N ______ CI Name ______

Item Affected (P/N) ______ Name ______

Next Higher Assembly P/N ______

Description of Change

Effectivity: S/N ___ Through S/N ___ Priority: ☐ Emergency ☐ Urgent ☐ Routine

☐ Production ☐ Rework ☐ Retrofit Model Affected: Engineering ☐

Remarks:

Prototype ☐

Qualification ☐

Operational ☐

Spares ☐

Class I ☐ Class II ☐

Accepted ☐ Rejected ☐

Originator:	Configuration Manager:	Engineering:
______ Date	______ Date	______ Date

Figure 35 Sample Change Request Form

Courtesy Wiley & Sons, Samaras & Czerwinski, *Fundamentals of Configuration Management*, Wiley-Interscience, 1971, p. 79

of this form is to allow manufacturing, inspection, or engineering personnel to recommend a proposed change based on practical considerations or problems arising during the production of the item depicted on the document. (A change request precedes a formal EO or CN.)

This form contains a description of the change, why it is needed and its impact on other areas such as associated hardware, manuals, specifications, and procedures. Information such as this is vital for an objective evaluation of a change. For example, the person originating the change may reduce the cost of manufacture by $10 per unit but the requested change may reduce its performance or reliability. Therefore, some thought by the requester should be applied to the impact of the change on other areas.

Another important area for the requester to evaluate is the total cost to implement the change. This includes production labor, material, fixtures, test equipment, changes in procedures, and engineering labor. The effects on the product's delivery date to the customer is also a key consideration. Saving a few thousand dollars in less labor and material may not offset a penalty for late delivery or customer dissatisfaction and future loss of sales.

A change request is rarely used by small companies. Instead, verbal discussion leads the preparation of a formal change document, which is prepared by the originator and submitted for approval. This saves an extra piece of paper and some work but reduces the thoroughness of the requester's analysis. If the CCB also fails to do its job, then the change's cost impact may not be determined until it's too late. Therefore, the use of a change request form is recommended for all changes, except those that are mandatory for the equipment or product to work reliably and safely. In this case, engineering and the CCB are responsible for recommending a more efficient method of implementation when one exists.

ENGINEERING ORDER: PRIME AUTHORITY FOR CHANGES

Engineering orders (EO's) are documents that describe and authorize a change to a released document. They are called many names in industry such as engineering change instruction, drawing change notice, engineering change order, and advance design change notice; however, their function is the same: temporary but binding supplementary documents to the released drawing. The EO is an integral though physically separate part of the basic drawing it modifies.

Engineering orders are prepared on A size, preprinted format sheets as shown in Figure 36. They contain all the vital identification data needed to assure correct association of the EO with its affected document. In addition, "was" and "is" conditions are described in the body of the form. After approval, the EO is issued to users of the basic document and stapled to it. Industry practice is to allow up to 5 EO's to be attached to a basic document before they must be incorporated into the original document. (However, EO's

☐ ENGINEERING ORDER
☐ CHANGE NOTICE

DOCUMENT NO.

EO/CN NO.

CHNG LTR

CLASS

ECP NO.

GO NO.

SH OF

EFFECTIVITY	FROM	THRU
PN		
CI PN		

DOCUMENT TITLE

NEXT ASSY

CI / SYSTEM

OTHER DOCUMENTS AFFECTED

APPROVAL	DATE
ORIG	
DFT	
CHECK	
ENG	
RELBL	
LOG	
CCB	

REASON FOR CHANGE

DISPOSITION OF PARTS		
IN PROCESS	IN STOCK	IN FIELD
CHANGE	CHANGE	CHANGE
NO CHANGE	NO CHANGE	NO CHANGE
SCRAP	SCRAP	OPTIONAL

OTHER

RI

DCR

PART CODE

RELEASE DATE	
EO	
CN	

DESCRIPTION OF CHANGE

Figure 36 Sample EO Form

should be incorporated whenever necessary for simplifying production usage or avoiding errors.)

Note that two blocks are shown in the upper left hand corner of Figure 36. When a change is to be incorporated immediately into the document vellum (original) the change notice (CN) block is checked instead of the EO block. This indicates that the change is not to be implemented by using a separate attachment to the print. The CN block is also used when outstanding EO's are excessive (e.g., 5 or over) and need to be incorporated into the original document and new prints issued to manufacturing and other users.

There are some well-defined rules for preparing EO's and CN's. These are summarized below:

1. Prepare one EO for each document affected by the change, even if all documents are interrelated.
2. Identify on the EO the numbers of other documents affected by the change such as PL's, schematics, and assembly drawings.
3. EO's are not revised, identified with a revision letter, and reissued. A new EO with a new number must be prepared canceling the preceding EO and describing the correct or new change. (This rule may be occasionally circumvented for minor record changes but not for configuration or design changes.)
4. EO's require engineering authorization before release.
5. The effectivity of the change in terms of the end item (configuration item) must be specified for critical changes.
6. EO's should be on 8½ by 11 inch sheets. Continuation sheets of the same size may be used when one sheet isn't enough. However, more than one continuation sheet should be avoided because a three sheet EO is difficult to work with.

ENGINEERING CHANGE PROPOSAL

Engineering change proposals (ECP) are formal requests for changes to documents or products already delivered to a customer. The government usually requires the use of an ECP to control engineering changes under its contracts. The ECP is normally prepared when a change is beyond the scope of the contract. That is, additional funds are needed, approved design or product specifications require alteration, or form, fit, and function are changed from the approved configuration.

The ECP is similar to a change request, except that it is usually more formal, detailed, wider in scope, and time consuming to process. Some data included within the ECP include:

- Description of change
- Justification of change
- Developmental requirements
- Alternative solutions
- Production and retrofit recommendations for cut-in points of change
- Effects on safety, reliability, service life, operating procedures, performance, weight, balance, training, maintenance, spare parts, publications, and computer programs.

When an EO is prepared in response to an approved ECP, the ECP number is recorded on the EO for easy cross referencing. ECP numbers are issued and controlled by one group to avoid duplication and omission of an ECP from the change document monitoring and control system within the company.

Note that while ECP's are normally a requirement of the government, private companies may find them useful on contracts for products that require tight configuration control. Thus, by adding the requirement for ECP's in the purchase order or contract, the supplier cannot make changes to the purchased equipment without first obtaining customer or company approval.

DEVIATION REQUESTS FOR LIMITED ALTERATIONS

During production of items, temporary departures from approved documentation may be necessary to expedite work or meet scheduled deliveries. A deviation request is prepared by engineering describing and justifying this departure (before it is actually made) for specific serial numbered products. When approved and released, the deviation form (see Figure 37) provides manufacturing with the authority to circumvent the official requirements for the equipment identified on the form. A copy of this form is kept with the product's manufacturing and inspection data package to explain and authorize the change in process, material, or construction taken from approved engineering data.

WAIVER REQUESTS FOR SPECIAL HARDWARE ALTERATIONS

A waiver request is similar to a deviation request. Its principal difference is in the timing of the request. For example, a deviation request is submitted before a change from approved practice is made by manufacturing. A waiver on the other hand is prepared after a change from approved practice is made. This change may have been done for schedule reasons or it may be a product of an error or deficiency (which while not fatal to the equipment is still undesirable as a standard practice). However, in either case the product does not conform to approved, released documentation. To allow manufacturing to ship the product as is, the customer or engineering may approve the change because

ABC Corporation Washington, D.C.	Code Identification 09876	**Deviation Request and Authorization**	Deviation Number
			Sheet Of
Project		Applied to Document	Date

☐ To be incorporated
or
☐ Limited as Follows
☐ Eng Model S/N____
☐ Qual S/N ________
☐ Proto S/N ________
☐ Prod S/N ________

Identification of Item or Operation Affected

Nomenclature ________________

Part Number ________________ Rev Ltr ____ Serial Number ________

Manufacturing Operation ________________ Mfr. order No. ____________

Test Operation Affected ________________

CI Part Number ________________ CI Name ____________

Deviation

Reason ________________

Description ________________

Approvals

Engineering ________________ Date ____

Quality Assurance ________________ Date ____

Test. Engineer ________________ Date ____

Configuration Manager ________________ Date ____

Project Manager ________________ Date ____

Customer ________________ Date ____

Figure 37 Sample Deviation Request and Authorization Form

Courtesy Wiley & Sons, Samaras & Czerwinski, *Fundamentals of Configuration Management*, Wiley-Interscience, 1971, p. 91

it does not materially affect the performance, reliability, safety, and operation of the product.

As with the deviation request and authorization, a copy of the approved waiver is filed with the manufacturing and inspection data package. The waiver format is similar to th᷊ of Figure 37.

OTHER CONTROL DOCUMENTS

A distribution list is an important part of the document control and change process. Prints of original or new documents and change documents must get to the right people to be effective. Therefore, reproduction must prepare and maintain a list of people who should have copies of all released documents. If a company operates on a project basis, the separate lists are needed for each project team.

A sign-out card is needed to obtain a vellum from Data Control for incorporation of a change notice or engineering order. It provides a formal request by Drafting to Data Control for a particular vellum and identifies the requester and date of release of the vellum to Drafting.

STOP ORDER FOR REDUCING WASTED EFFORT

A stop order is prepared by engineering to instruct manufacturing, purchasing, or testing to stop work on a product because a change is in process that will invalidate the work being done. Figure 38 shows a stop order. It is uniquely identified as are EO's and waivers, and it contains identification data to assure that the order is clearly applied against a particular item(s). An estimated lift date for the order is also recorded to allow manufacturing and other departments to plan for the change.

Because of their critical nature, stop orders should be released within a day to avoid undue costs in wasted labor and materials.

6.3 Revision of Engineering Documents

Incorporation of changes into the original document requires strict adherence to established practices to minimize errors and incorrect procedures.

> **CASE IN POINT: A company that allowed careless documenting of changes found that it needed to refer to its historical data to determine when an invention had been first recorded. The development of the invention was traced to a drawing that carried a specific identifying number. However,**

Sheet — Of —

Estimated Date of Stop Order Lift: ________	**Stop Order**	SO Number ________
CI Serial Numbers Affected		Date ________

Part Number	Item Name	Next Assembly Part Number	CI Part Number

Reason for Stop:

Engineering Notes:

Responsible Engineer	Project Engineer
Other	Other

Manufacturing Notes:

Manufacturing Representative	Configuration Manager
	Project Office

Figure 38 Sample Stop Order

Courtesy Wiley & Sons, Samaras & Czerwinski, *Fundamentals of Configuration Management*, Wiley-Interscience, 1971, p. 94

when the original was located, the company discovered that the identifying number had been erased and a new number inserted. While the new identifying number had been recorded in the drafting room's log book, the drawing was useless as a record because a major change had been made in the drawing itself, and this change had not been described on the drawing, dated, or signed by the draftsman.[16]

The first change incorporated into a vellum is identified with an A in the "LTR" column located in the upper right hand corner of the drawing. (See Figure 39.) The change is described in the next column and the date and approval signatures are placed in the next two columns. Approval signatures should include the checker and responsible engineer. Note that the second revision block example references the revision notice or change document number for a description of the change rather than describing in that column.

Some companies cross reference the change in the revision block to the drawing field as shown in Figure 39. Note the circled B1. This indicates the first part of a change covered by the B revision to the drawing. B2, B3, and B4, are also identified for ease of locating the changes covered in the revision block and identified by (1), (2), etc.

Note that the left hand zone block is used for large drawings to simplify locating a change. (See Chapter 4.) It is not used on this drawing because of its small size and the use of circled cross referenced numbers.

Changes to drawings can be made by erasure, crossing out, or by redrawing the document. When incorrect data are crossed out, don't obliterate the information. It should still be readable for future reference.

Rules for revising document vellums are:

1. Changes incorporated at one time are identified by the same revision letter.
2. Revision letters are always capitals and exclude I, O, Q, and X. (See Chapter 5.)
3. If a symbol is not used to locate a change on the field of drawing, the zone column should be filled in.
4. The approval block should be signed by the responsible engineer and checker (if used) or draftsman if no checker is used.
5. On multiple sheet documents, the top sheet and those sheets affected by the change are reidentified with the applicable revision letter. Unaltered sheets retain their last revision letter. (Many companies change all sheets.)
6. When a new sheet is added, the note: "This sheet added" is recorded in the revision block of the new sheet.
7. The revision block should contain reference to enable user to check back on complete reason and authorization for change.

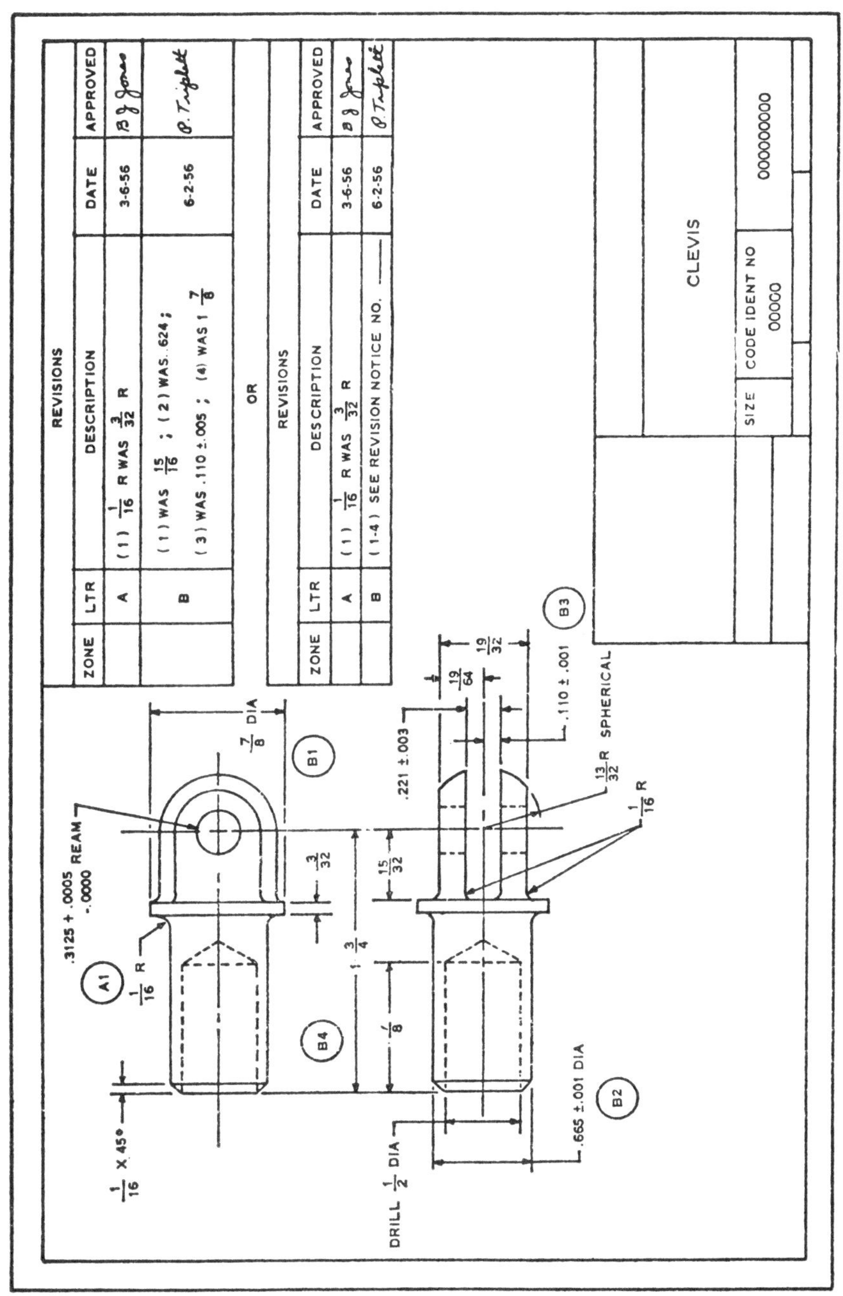

Figure 39 Recording Revision

When a document is redrawn because of a change, the next revision letter is added in the revision block. In addition, the note: "Replaces revision X with change" is added to the description column in the revision block of the new document.

If a drawing is redrawn without changes, the revision letter of the old drawing is used. The following note should be added near the title block:

"Drawing 804XXXXX, Rev. B (6/19/74) redrawn *without* change."

Approval signatures are needed on a redrawn drawing as were recorded on the old one.

A superseded document contains the word "SUPERSEDED" over the title block and the note:

"Replaced with change by Revision X."

RECORD CHANGES

A record change revision does not alter the product configuration, design, manufacturing processes, or materials. Record changes usually do not need an EO/CN for Drafting to incorporate the change.

Criteria for determining whether a change is a record change or not are:

- Does it affect form, fit, or function?
- Does it affect technical notes, processes, and reference documents?
- Does it affect other departments/functions? For example, is the new information needed by Publications, Purchasing, Manufacturing, Inspection, Product Assurance, Engineering, or Testing?

If the answers to any of the above questions are yes, an EO/CN is needed to make the change. Examples of record changes are: (1) add arrowheads, (2) darken lines, (3) correct misspelling, and (4) add application data, e.g., next assembly and used-on data.

6.4 Obsoleting Data To Prevent Inadvertent Use of Old Designs

The purpose of obsoleting a document is to:

- Remove it from usage
- Eliminate it from active files
- Notify users that it is no longer a valid document

To remove released engineering data from the active data control files requires a special procedure that ensures a consistent and well-defined system. The exact method is not critical as long as a formal system is defined, documented, and understood by data users.

If a new document has superseded the obsolete one, the release form and other associated records should contain the note: "Supersedes document number XXXXXXXX, March X, 197X."

6.5 Dual and Multiple Release

Dual release refers to having more than one place in the organization where a new document can be released. This is not a good practice because duplicate record-keeping and filing are needed and the release system is weakened because of possible differences in practices between the two groups. In addition, engineers may learn to manipulate the system to their seeming advantage. That is, they may release an incomplete document or change notice through one lenient release group when the other is more stringent and demands that all requirements be met before it distributes a document.

Multiple release refers to having more than one version of a document in active files. That is, all documents used to build hardware are kept in the print file so that they are available to use when needed to repair or maintain the product. For example, equipment may have been shipped during the last year built to revisions A, B, and C of a drawing. Therefore, prints should be readily available to the staff when the configuration of a particular equipment is needed.

CASE IN POINT: At one small company, it was found that an obsolete drawing was really better than a newer one for a new application. The obsolete drawing was retrived from the inactive files and used for the new product application. The cost of storing and retrieving the old drawing was a fraction of the cost of redesigning and testing a new item to do the same job.

6.6 Field Alterations: Don't Neglect Their Documentation

Changes are often made to products in service or already delivered. If accurate identification of the configuration is desired, a careful system of documenting changes is necessary. The critical steps in a tightly controlled system are: (1) engineering issues retrofit notice and description, (2) change implementer installs modification kit or new parts and identifies equipment as having been altered. This identification may be by recording what was done in an equipment log book or by adding a new identification or a modification plate to the equipment, (3) submit a retrofit completion report to engineering and service manager, and (4) service center adds retrofit report in equipment's data file.

7

7.0 Records, Reports, and Microform: Vital Administrative and Information Tools

Records and reports are essential parts of all modern organizations. The D&D department is not immune to this need. But it has its special requirements as were briefly discussed in Chapters 2 and 3. These requirements are covered in the following sections. However, before proceeding, definitions of records and reports follow:

Record: A record is a document that is used to set down in permanent form an indication of what actions or events have occurred. It preserves an account of what has transpired during a job or program.

Report: A report gives an account of activities at regular intervals. It may also give a formal presentation of facts or a summary of events on a one-time basis.

Note that while the terms "record" and "report" have specific meanings, the documents described in this chapter usually serve both functions of providing historical information and data on the status of operations.

BENEFITS AND APPLICABILITY OF THIS CHAPTER

Records and reports are a subject that create bad feelings in technical people because they steal time from creative and productive efforts, are boring and routine, and often don't seem to have worthwhile purposes. There's certainly truth to all these objections, but the modern day world would collapse without adequate record keeping and report preparation. Specific benefits of these documents are (1) provide historical data for reference and management decision making, (2) provide legal data, (3) are a medium for information exchange, (4) help identify problem areas and corrective actions needed, and (5) assure management that a defective or dangerous part has been replaced.

The benefit of this chapter is to help technical people understand why record and report activities are important in the modern day world and to give examples of specialized records and reports related to the D&D operation. (Conventional and management reports are covered in Chapters 2 and 10.) Another value of this chapter is its discussion of microform technology and its value to engineering data activities.

An example of how good records help management and engineering make better decisions and improve the company's competitive edge follows:

> **CASE IN POINT: One company found from its records of drawing production (square feet of drawings produced per year) that its output increased by 15%. A check of the previous year's output also revealed an improvement of 7%. Thus, based on the greater productivity of the drafting group, it was able to reduce its drafting bids and the company could submit more competitive prices for its services and its products. Without this information, the company would be at a loss to adjust its drafting costs up or down. In addition it would have no objective way for evaluating the effectiveness of new D&D techniques for reducing costs or improving output.**

When life and products were simpler, a designer or engineer could keep much of the information needed in his head, but today the complexity of engineering requires good records and reports. Without these documents, important decisions or events are soon forgotten or distorted. However, requirements for records and reports should not be accepted blindly. If a document seems unneeded, redundant, or impractical, D&D personnel should question their supervisors about its merit; there are certainly many cases where records and reports can be eliminated or streamlined.

> **CASE IN POINT: At Hoffman Electronics Corporation, it was found that two release forms were being used to do the same job. One form was a traditional release form used for many years. The other form was relatively new and had been created for inputting data to a new computer data base. The two forms were used for about 1 year until someone noticed that, with minor changes, only one form need be filled out to**

do the job of two. The second form was discontinued saving the company about $1000 per year in reduced clerical labor, less filing space, and less paper.

Note that because all companies require records and reports, the information presented in this chapter applies to all types of companies and D&D operations.

7.1 Status Cards and Log Books: Records of Critical Events

Drawing status cards and log books were covered in Section 3.3. Note that these are critical documents that help prevent chaos in the D&D operation. They inform the D&D personnel as well as managers and engineers as to where a particular document is in the preparation cycle and its legal status: Is it released by engineering for production use? What is the latest revision letter? Is it being changed by someone now? The card shown in Section 3.3 is only one part of the status system. The second part, which comes before the release record, is the inprocess status card or log book. This document contains the location of each document in the D&D department. For example, it tells the person looking for it whether it is in design, drafting, check, signoff, or configuration control board. For smaller companies, this may not be considered an essential requirement, but for large companies, this kind of status information can help improve scheduling, coordination, and efficiency. (See Chapter 10.)

Drawing number log books are a necessity for all companies. An accurate and up-to-date record of all drawing or change document assignments must be maintained by data control or drafting to avoid duplication of number assignments and to enable tracking the status of documents that have never been submitted to drafting from design or engineering.

7.2 Indentured Drawing List

An indentured drawing list (IDL) provides valuable information on the makeup of a product design. It lists all the drawings that are needed and their interrelationships in terms of higher and lower assemblies.

Various methods exist for identifying the indenture level. These were discussed in Section 4.2.

Besides overall data on the configuration of the product, IDL's help in tracking progress on the job in terms of what has been released as of a certain date, the various types of drawings that will be prepared, and how many drawings will be needed. In other words, they help plan work better and define what is required. Another important use of the IDL is in allowing identification of all drawings needed to begin fabrication of a particular assembly. Thus, engineering or manufacturing can request that D&D concentrate on preparing drawings for a particular assembly.

IDL's should be prepared as early as possible in the product design and development effort so that a baseline for tracking and correcting drawing requirements can be available to all departments involved in the effort. Of course, they should be updated periodically by engineering or design to ensure as accurate a status as possible.

7.3 Drawing and Change Status List For Tracking Data Changes

A drawing and change status list can be used by D&D to keep track of each drawing and its associated changes. Figure 40 shows a sample form used to identify each document, its revision letter status, and EO status. This list

ABC Corporation Washington, D C | Code identification number 09876

Configuration Identification and Accounting Index—EO Status

Number | Rev

Project ________ Date ________ Superseding ________ Sheet of

EO Number	Title	Type[a]	Drawing Number[b]	Part Number Affected	CI Name	Effectivity		Related Data Affected	Change Identification Number	EO Release Date
						From	Through			
	[a] Type M - Mandatory I - Improvement R - Record X - Expedite DV - Deviation WV - Waiver SO - Stop order									

[b] An additional column for the next higher assembly could be added if desired

Figure 40 Configuration Identification and Accounting Index

Courtesy Wiley & Sons, Samaras & Czerwinski, *Fundamentals of Configuration Management*, Wiley-Interscience, 1971, p. 242

provides a current configuration record for the product and allows D&D personnel to know how many EO's have been released against a particular drawing and how many are currently in process. Thus, if a complete picture for a drawing is needed the latest revision and associated amendments can be readily obtained using this list.

CASE IN POINT: Failure to provide a cross-reference index between documents and change orders can lead to lost time, confusion, and lost profits. For example, a medium-size manufacturing company found that failure to keep a cross-reference index for documents and changes in process against them, cost thousands of dollars a year because engineers and designers often created EO/CN's that were already in preparation by another engineer, in conflict with other in-process EO's, or being cancelled by other EO's.

With an average of 50 duplicate or unnecessary EO's processed each year, the extra design, engineering, drafting, check, and support labor came to an unnecessary additional cost of over $5000 per year.

A computer-type report is shown in Figure 41. This report supplies the following data for each document in alphanumeric order: (Read columns from left to right in Figure 41.)

- Release status (R means released)
- Document number
- Revision status (number represents an advance release while a letter is a production release)
- Description or brief title of document
- Size and number of sheets
- Current or latest revision or change status (number represents revision status of an advance release while, E502XX represents an EO against the document, A C502XX represents a change notice. If an XX follows an EO or CN number, this means that the document is pending and has not been released yet). Letters under the EO/CN column and to left of comma give revision letter of drawing associated with parts list identified under DWG/DOC column.
- The last date an action was taken against document file.
- The remaining columns identify the type of action last taken against the document; e.g., new release, change action, new usage, deletion of usage, or stop order

CURRENT STATUS REPORT   EPLT/CDLT   REPORT 290-17   05/15/   PAGE 581

REL	DWG/DOC NO.	REV	DESCRIPTICN	SIZE	SHTS	CURRENT EO/CN	LAST ACTIVE	NEW REL	CHG ACT	NEW USG	DEL USG	STOP ORDER
D	PL8855400051-1					C A,	09/27/					
R	8855400052	B	MICROELECTRCNIC CEV	A	20	C51783 B ,	09/05/					
D	PL8855400052-1					C A ,	01/12/					
R	8855400054		MICROELECTRONIC DEV	A	21	C ,	04/03/					
D	PL8855400054-1					C ,	07/07/					
R	8855400056	A	MICROELECTRCNIC DEV	A	10	C20373 A ,	01/12/					
D	PL8855400056-1					C A ,	01/12/					
R	8855400057	B	MICROELECTRCNIC CEV	A	17	C51784 B ,	09/24/					
D	PL8855400057-1	A				C A B,	09/25/					
R	8855400058	9	MICROELECTRCNIC DFV	A	18	C51785 B ,	09/11/					
D	PL8855400058-1					C ,	09/18/					
R	8855400059	A	MICROELECTRCNIC DEV	A	20	C20286 A ,	01/12/					
C	PL8855400059-1	A				C A ,	08/04/					
R	8855400061	C	MICROELECTRCNIC DFV	A	12	C51786 C ,	09/17/					
D	PL8855400061-1	A				C A ,	07/14/					
R	8855400063	B	MICROELECTRCNIC DEV	A	10	C51787 B ,	10/10/					
D	PL8855400063-1	A				C A B,	01/12/					
D	PL8855400063-2	A				C A B,	01/12/					
R	8855400064	A	MICROELECTRCNIC DEV	A	10	C20292 A ,	01/12/					
D	PL8855400064-1	A				C A ,	07/25/					
R	8855400065	D	MICROELECTRCNIC CEV	A	13	C51788 D ,	09/24/					
D	PL8855400065-1					C B ,	10/25/					
R	8855400067		MICROELECTRCNIC DEV	A	10	C ,	04/03/					
D	PL8855400067-1					C ,	07/20/					
R	8855400068	C	MICROELECTRCNIC DEV	A	11	C27520 C ,C45972 XX	02/14/					
	8855400068-1					02729 STOP ORDER	08/11/					C6/25/
D	PL8855400068-1	A				C A ,	0?/24/					
D	PL8855400068-2	A				C A ,	06/04/					
D	PL8855400068-3	A				C A ,	06/04/					
D	PL8855400068-4	A				C A ,	06/04/					
D	PL8855400068-5	A				C A ,	10/26/					
D	PL8855400068-6	A				C A ,	06/04/					
C	PL8855400068-7	A				C A ,	09/15/					
D	PL8855400068-8	A				C A ,	09/15/					
D	PL8855400068-9	A				C A ,	09/15/					

Figure 41 Current Status Report

Selection of a manual or computer report depends on the volume of drawings and changes processed every year. The cost of data processing services versus manual preparation needs to be evaluated systematically and formally before making a decision to computerize document status reports.

7.4 EO and ECP Status Record: Helps Know Where an EO/ECP Is in Processing Cycle

When numerous Engineering Orders (EO's) and engineering change proposals are being prepared by the staff, a detailed tracking list may be needed to know where each change document is in its preparation cycle. (For a medium size company, the number of EO's and ECP's prepared in a month may exceed 500.) This tracking list/record can be used both as a means for logging out or assigning numbers to EO's and as a basis for identifying where a particular EO is in the processing cycle.

A similar record can be maintained for ECP's. An ECP is a formal change proposal submitted to the customer when he needs approval of engineering changes (see Chapter 6) before they can be implemented and when additional funds are needed to implement the change. The requirement for approval of changes by the customer ensures his ability to control the product configuration and to prevent unauthorized changes that could have bad effects. For companies doing government work, the procedure for preparing ECP's is described in MIL-STD-480, Engineering Changes, Waivers, and Deviations.

7.5 Configuration Identification and Accounting Index Defines Product Configuration and Status of Approved Changes

Some companies require careful accounting of changes and their effects on hardware. Information such as new part numbers, effectivity, and other documents affected by a change are recorded in a summary report that allows engineering and management to know what the product configuration is for various serial number end items delivered to the customer, or still on the production line. Key data elements include:

- Change document number and title
- Effectivity (retrofit and production sequences)
- Change submittal and approval dates
- Identification of spares affected by change
- Date change was incorporated into hardware
- Part numbers affected and new part numbers created as a result of change

Although D&D is not usually required to prepare this report, the staff should be familiar with its purpose and existence for possible reference during problem areas requiring determination of document and change status information.

7.6 As-Built List: Tells What Was Shipped To Customer

An as-built list is a record that identifies the product configuration that was shipped to a customer. This record allows positive tracking of lower level assemblies when problems occur or when information is needed as to what part numbers were used for various serial numbered equipment delivered to customers.

Information recorded on the as-built list includes:

- Assembly level identifer
- Part number
- Part nomenclature description
- Next assembly number
- Acceptable revision letter of drawing used to build part
- Revision letter of drawing part was built to
- Serial number of part used in product

The value of the as-built list is in the following information it provides various departments within the company:

1. Informs maintenance and repair people of the constituents of a particular equipment or facility.
2. Allows determination for each equipment as to whether a new part was installed before placement in stores or before shipment to the customer. Therefore, it can prevent unnecessary work in returning equipment to manufacturing to determine if it has the right parts or assemblies.
3. Provides publications with data for providing operation and maintenance manuals with the correct configuration of each serial numbered product.
4. The list can also be used to identify changes made to the equipment after shipment by field personnel. Thus, an accurate record of the equipment's configuration is always with the equipment. However, this change information must be sent back to the plant so that quality assurance and manufacturing can update their as-built list.

Whether an as-built list is needed depends on company and customer requirements. Management may decide that the cost of preparing and maintaining this list does not result in advantages that offset the cost. However, many companies have found through bad experiences that an as-built list is worth the investment. A bad experience follows:

CASE IN POINT: One company found that a group assemblies used in their products contained defective parts that created safety hazards. Because the company didn't use as-built lists, it had to inspect all products at their locations to find the defective subassemblies. Since hundreds of these products had been delivered, the company had to spend thousands of dollars to physically check each product for the defective assemblies. Availability of an as-built list at the company would have only required inspection of the lists for the location of the desired serial numbered assemblies.

7.7 Preferred Parts Reports for Promoting Lower Costs

The Standards department is responsible for assuring that reliable but low cost parts are used for the company's products. It is also responsible for ensuring that customer requirements are met for special products needing highly reliable and durable parts. To assure these objectives, Standards prepares reports that identify and define those parts that are preferred for all company products. These reports must be studied by designers and checkers to ensure that designs include only acceptable parts. If non-preferred parts are needed, special requirements must be applied to assure their acceptability to the customer or company.

7.8 Checkprints and Obsolete Documents For Future Reference and Resolution of Conflicts

Checkprints and obsolete engineering documents are records that need to be stored for possible future reference. Marked-up prints from engineering and checkprints may be stored in the D&D department for 6 to 12 months, depending on department policy. Obsolete documents are usually kept in the data control areas for a longer time; e.g., 2 years.

The purpose of keeping these documents is to correct errors made in translating information from the checkprint or old drawing and not caught until a month or more later. They also provide a record of disagreements between checker and engineer and help avoid false accusations when things go wrong. Another reason for keeping files of old documents is that they can provide evidence of when inventions or proprietary developments were first made in case legal actions between the company and outside agencies or other companies are required.

7.9 Microforms for Ease of Data Retrieval and Space Savings

Any discussion of records and reports is incomplete without a review of microform techniques used in industry to reduce retrieval time, storage space, and operating costs. Many of the documents discussed in the preceding sections are amenable to microform conversion. However, a prime application for the use of reduced image documents is in the area of production documents such as drawings and specifications. Microform documents can be effectively used by D&D, engineering, manufacturing, quality assurance, and customers.

Before proceeding with a more detailed analysis of microform technology, a summary of key definitions is presented:

Aperture Card: an electronic accounting business card containing a microform image mounted in a window cut out of the card.

Blowback: creation of normal size print from a reduced image or microform.

Frame: a space allotted for one image on a film strip.

Hard Copy: a normal size copy produced from a microform image.

Microfiche: sheet type film containing rows and columns of microimages.

Microfilm: conventional roll photographic film containing a sequence of microimages on it. Film sizes are 8, 16, and 35 millimeters.

BENEFITS OF MICROFORM

Many benefits from microform are available to small and large companies, depending on their special needs. However, its chief advantage is its compact storage of data; e.g., 98 standard size pages can be recorded on a 4 by 6 inch transparency and sent anywhere in the country for 10¢. When it gets there, it can be blown up to full size and read. Some other specific benefits are:

—Much smaller space required for data files
—Cheaper to print than a normal size drawing print
—Easier to change files requiring constant updating
—Reduces need for extra prints as backup to meet requests for copies.
—Reduces need for supplying normal size prints when a designer or draftsman wants only a brief reference to the document for a small amount of information.

—Requires smaller disaster vault for storage of critical documents. Besides requiring less space, the cost of a large, well-built vault can get very high.

CASE IN POINT: GTE Sylvania Incorporated (subsidiary of General Telephone and Electronics Corporation) reduced its record-keeping and duplicating manpower requirements by 77 percent by reducing their material audit files and engineering documentation to microfilm.[17]

CASE IN POINT: Eaton Marshall Division of Eaton Yale & Towne, Inc. found that it could save 88.5¢ per drawing in the area of reproduction and distribution of engineering drawings. Over a year, a $10,000 reduction in cost for an output of 5200 new and revised drawings was realized by microfilming its drawings.[18]

REVIEW OF MICROFORM CHARACTERISTICS

The various types of microform used for documentation storage, retrieval, and use are described next. Note, however, that some forms are not practical for D&D or engineering data. Figure 42 shows sample microforms and equipment used in industry.

ROLL MICROFILM: This type of microform is excellent for reducing file requirements for letter size documents. About 2000 standard size documents can be recorded on a single roll of 16 mm, 100-foot long film. Rapid data retrieval and printouts are readily obtained from motorized viewers and hard copy printers. Container types include cassettes, magazines, and cartridges.

The advantages of this microform are reduced storage space, elimination of incorrect filing, and rapid retrieval of data. Its key disadvantage is that updating files is difficult because all the documents on the film would have to be refilmed.

JACKET MICROFILM: Jackets are a unitized type of microform which consist of strips of 16 or 35 mm film inserted in a clear, acetate carrier. This type of microform is used for records that require frequent updating. Up to 75 standard size documents can be contained on a 4 by 6 inch carrier sheet.

The advantages of these jackets are easy revision, fast retrieval, and ease of duplication. Their disadvantage is that hard copy printout is slow if the entire sheet must be duplicated.

APERTURE CARD: 35 mm aperture cards (standard electronic data proces-

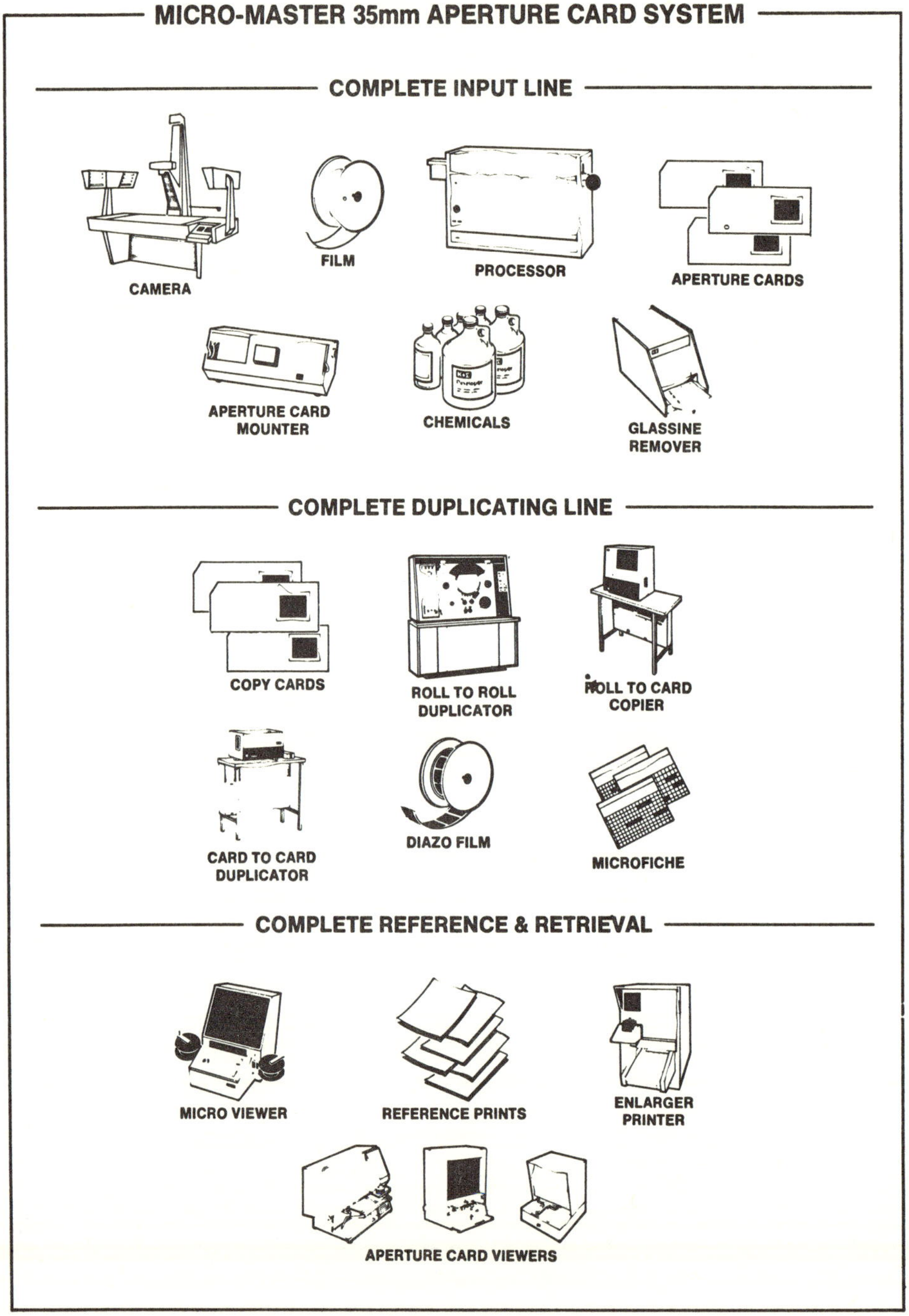

MICRO-MASTER 35mm APERTURE CARD SYSTEM

COMPLETE INPUT LINE

CAMERA

FILM

PROCESSOR

APERTURE CARDS

APERTURE CARD MOUNTER

CHEMICALS

GLASSINE REMOVER

COMPLETE DUPLICATING LINE

COPY CARDS

ROLL TO ROLL DUPLICATOR

ROLL TO CARD COPIER

CARD TO CARD DUPLICATOR

DIAZO FILM

MICROFICHE

COMPLETE REFERENCE & RETRIEVAL

MICRO VIEWER

REFERENCE PRINTS

ENLARGER PRINTER

APERTURE CARD VIEWERS

Figure 42 Microform Equipment and Types

Courtesy of K&E Co.
20 Wippany Road
Morristown, N.J. 07960
Source: *Engineering Graphics* Ad, Sept., 1973, p. 8-9

sing cards) are commonly used for engineering data release systems, product maintenance drawings, and data submittals to customer. A microfilm frame is cut to size and pressed onto an adhesive strip that surrounds the card window as shown in Figure 42. Card dimensions are 3¼ by 7⅜ inches and information as to the document name, number, and size is typed along the top edge of the card.

The advantages of this microform type are fast retrieval, reduced storage space, easy duplication, machine sortable, easy to up-date files, and convenient to use by staff. The main disadvantage of aperture cards is that full size enlargement of large drawings is not practical.

MICROFICHE: Microfiche (sheet microfilm) is used primarily for book-type documents such as specifications and standards. It contains multiple images arranged in rows and columns. Titles and other administrative data are recorded on the top of the sheet for easy identification without special equipment. Many pages of information, e.g., 98, can be recorded on one 4 by 6 inch sheet of microfiche.

Advantages of this microform type include ability to record an entire document such as a procedure or specification onto one or two sheets of microfiche, fast retrieval, and low duplication cost. A principal disadvantage is difficulty in revising data on the sheet.

105 MM MICROFILM: This large sheet microfilm is used when high quality enlargements/prints are required. It contains identification information in eye-readable form at the top of the sheet.

Advantages of this microform are fast retrieval, reduced storage space, reduced clerical labor, and full size print blowbacks. Main disadvantage is it is larger than other types of microform.

COMPUTER OUTPUT MICROFILM (COM): COM is a technique for transforming data processed by a computer into human readable form. The output of the computer is converted from digital to human readable form and projected onto microfilm at a high rate (e.g., 250,000 characters per second). Output is either on roll microfilm or microfiche.

The advantages of COM are lower cost, faster duplicating speeds, elimination of many paper reports, and space reduction. Its disadvantage is no paper copy for making notes.

MICROFORM EQUIPMENT

A variety of equipment and materials is needed to produce, store, retrieve, and reproduce microforms. Figure 42 illustrates some of the typical equipment required.

CAMERA: A planetary camera is needed to create good quality microfilm. This equipment consists of a camera that can be moved up or down a vertical column and copy plate for setting up the original document to be photographed.

PROCESSOR: A processor develops, fixes, washes, and dries microfilm. Most processors are daylight working and do not need darkroom facilities.

HAND MOUNTER: A hand mounter converts roll film to aperture cards. (May be done automatically by the camera.)

DENSITOMETER: A densitometer checks the exposure and density of film negatives to assure acceptable quality.

DUPLICATOR: Different types of duplicators are used for reproducing roll film, aperture cards, and microfiche. (See Figure 42 for examples.)

READER: Various types of readers are used to present an enlarged view of the microform of interest. The enlarged image is projected onto a large screen (e.g., 18 by 24 inches) to allow the user to study the information required.

READER-PRINTER: The reader-printer provides both blowback images and hard copies of microform images. This device has controls for selecting the desired image from a film or microfiche source and for making prints.

SAMPLE STORAGE AND RETRIEVAL SYSTEM

Various types of storage and retrieval systems are available using microforms. The nature and scope of these systems depend on the needs and finances of each company using them. For example, a company may need an information system that combines microfilm, automated retrieval mechanisms, and video display units.

Information is stored in one place on computer memory, aperture cards, and microfiche. Hundreds of thousands of these records never leave the storage area but the engineer or draftsman can get a remote display or print of any drawing in the system.

The requester has access to a keyboard for identifying and requesting the drawing needed and directing the system as to which kind of output is needed: video display, reduced immediate printout copy, or delayed (because of manual reproduction) full-size print. The user has control over the information presentation. He can scan each sheet of a document or move to different areas on large drawings. He can also enlarge selected portions of the drawing for greater detail.

Some of the advantages of this system are:

- Automatic retrieval of data in seconds
- Original data don't leave storage area

- Unnecessary prints are avoided
- Allows centralization and control of all data in one place
- Eliminates walking to print center and waiting for a print that may not be needed
- Eliminates misfiling, lost records, and the usual labor needed to manually retrieve documents

7.10 Special Microfilm Requirements for Preparing Engineering Data

Documents to be microfilmed must be prepared to more stringent standards to produce high quality blowback prints. Therefore, a company planning to implement a microfilming program should be sure that the drafting department is aware of the following requirements:

- Lettering must be uniformly spaced and consistent in weight and style
- Lettering must be at least ⅛ inch high
- Vertical-style lettering without serifs is preferred
- The space between lines of lettering must be at least half the letter height
- Addition of changes to original drawing must conform to the style of the original draftsman
- Lines must be sharp, even, dark, and uniform in intensity
- Lines should be heavier than for normal drawings
- No two lines shall be closer than .06 inch
- Overcrowding of drawing must be avoided
- Select a scale so that adequate space exists between lines on the drawing
- The number of standard drawing sizes should be kept to a minimum.
- Whenever possible typewritten or preprinted letters should be used instead of free-hand letters.
- Use a horizontal bar between the numerator and denominator in all fractions.

8.0 Reproduction of Drawings and Data: Getting Technical Information to Users

Reproduction of engineering drawings and data is a constant, permanent, and key activity in most organizations. While it is often a suboperation of a larger group such as design and drafting or data control, the reproduction facility plays an important role in the efficient operation of a project or an entire company. An effective reproduction operation is characterized by the following features:

* Rapid supply of quality prints to engineering, production, procurement, testing, and quality assurance.
* Low cost copies of required documents.
* Low risk of original document damage or loss.
* Capability to respond to sharp demands in output.

The reproduction operation can be viewed as a separate cost center for evaluating its efficiency and profitability. When treated as an integral part of a larger operation, a poorly managed and high cost reproduction group can often escape detection by management. Thus, it can continue to adversely affect the overall company profitability. Although the reproduction facility need not be an independent operation, it can still be evaluated as if it were one to improve the quality and efficiency of its services.

BENEFITS AND APPLICABILITY OF THIS CHAPTER

This chapter presents an overview of the reproduction of engineering data. Specific benefits available from the material presented are:

- Gives guidelines for selecting a reproduction system.
- Provides a variety of reproduction methods available.
- Gives requirements for a reproduction facility; e.g., size, layout, and equipment or materials.
- Presents a review of special requirements to consider before selecting a system.
- Provides a review of reproduction materials, their characteristics, and costs.
- Gives cost savings techniques.
- Provides guidance in avoiding problem areas.

Every technical company needs a reproduction system. The equipment selected depends on its needs and financial situation. Therefore, while most of the equipment described in this chapter is applicable to all size companies, certain items will not be practicable for small companies. Refer to Table II in Chapter 1 to evaluate the cost effectiveness of any system that seems too sophisticated for your company.

8.1 What To Consider for an Effective Reproduction System

Analysis of reproduction requirements is often ignored by management and engineering and delegated to someone with experience in the reproduction area. However, while this person may have expertise in specialized reproduction processes, he is not in an ideal position to design the best system. For example, he doesn't know what the most important reproduction needs and objectives are from the total company viewpoint. Consequently, it is necessary that he be given guidelines in terms of what the immediate and long term reproduction needs of the company are. These guidelines can be provided by management after it has analyzed the following requirements:

- Quality (minimum standards for density and contrast of image lines; paper weight and texture)
- Capacity (number of copies per month required)
- Speed (elapsed time acceptable to produce a certain number of copies)
- Cost (maximum acceptable cost per copy, including labor, equipment, space, and supplies)
- Reliability (maximum acceptable breakdowns per month)
- Safety (acceptable risk in terms of injury to personnel or damage to originals)
- Convenience
- Ease of installation and maintenance (also speed with which the equipment can be repaired after a malfunction)
- Durability of copies and reproducibles (how long must copies remain useful

before fading or other deterioration requires new copies?)
- Growth potential
- Versatility

A systems or total approach towards these requirements is useful in attaining the most effective reproduction operation. For example, while some requirements are compatible and do not require tradeoffs to achieve the most effective system, some requirements compel sacrificing desirable qualities to fall within prescribed limits of cost, space, and flexibility. Therefore, the selection of one requirement at the sacrifice of another must be made on the basis of which one enhances the effectivity of the whole operation, including its users.

Most of the preceding requirements are obvious in terms of what they imply. However, growth potential and versatility need further discussion. "Growth potential" refers to the ease with which the reproduction operation can grow in absolute capacity or output, efficiency and lowered costs, and new capabilities. Consequently, a good reproduction system is designed to allow for more equipment, personnel, and new functions with minimum cost and disturbance to the operation of the system. Thus, while the needs of the present system do not call for equipment to convert microfilm or aperture cards into hard copy, the power and space requirements should be available if this function is a possible requirement in the near future.

"Versatility" refers to the ability of the system to handle a variety of reproduction tasks. For example, a versatile system can reproduce opaque or translucent originals varying from standard size sheets to roll size; it can produce copies on white or colored paper, projection film, or card stock; it can readily produce anywhere from 1 to hundreds of copies conveniently; and it can bind copies into booklet form.

In evaluating an existing or proposed reproduction system, it is useful to prepare a priority list before the specific system is studied. This allows a more objective analysis to be made by providing a baseline for comparing the characteristics of the reproduction system with what is needed.

In summary, a clear-cut identification and priority listing of reproduction system requirements is essential for satisfying the company's needs at minimum cost and best performance. This can be achieved by listing the preceding requirements in the order of their importance or by assigning a weighting function (such as 1 through 10) next to each requirement to indicate the relative attention that it should be given.

8.2 Reproduction Cycle

The reproduction cycle is often written off by managers and engineers as a trivial operation. However, the many steps involved, while not complicated, do add up to a significant labor expense when multiplied by thousands of repetitions each year. Figure 9 shows the operations that must be performed to produce one copy of an original document. Of course, for a small company, this system can be made less complex and formal, but as a company grows or requires greater control, the cycle usually develops into the one shown in the Figure.

To ensure a smooth reproduction cycle, a priority system is needed for duplicating originals. The normal approach is first-in, first-out. However, special conditions or projects may need to violate the standard priority system. Thus, the reproduction supervisor should be notified in writing by management that certain projects can have their documents copied before others when specifically requested. The duration of this higher priority should also be stated. As a general guideline, the following standard priority system may be used when the reproduction operation cannot satisfy all staff demands for copy quantity and turn-around time.

Priority	Project/Document Type
1 (top)	Proposals/new business reports
2	Production documents (e.g., build-to drawings, parts lists, and wire lists)
3	Correspondence
4	Reports, specifications, and manuals to customers
5	Design and engineering documents
6	Research and development data on new products
7	Memos, technical data, and miscellaneous

Another aspect of the reproduction cycle is keeping records for assigning costs to various departments in the company. A reproduction request can be used to initiate the reproduction process and to provide a record for determining monthly copying work. At the end of each month, the reproduction work done can be summarized on a sheet and costs allocated to each department based on an average charge per copy or per square foot.

8.3 Facilities and Equipment

The reproduction operation requires special facilities and equipment. These include copying/printing machines, work tables, binding equipment, storage cabinets, segregated work areas and file cabinets. Inadequate facilities and equipment can reduce an experienced reproduction team to an ineffectual operation. Therefore, special care is required to assure adequate room to work in and the right quantity and types of equipment and materials.

TYPES OF FACILITIES

Two basic types of facilities exist: open and closed access. The open access facility allows anyone in the engineering department or company to use the reproduction equipment. This is the common method for small companies. However, as the company grows, problems arise because of conflicts in priority of use, long waiting lines, misuse of the equipment, and inadequate reordering of operating supplies and materials.

The closed or restricted access facility is partitioned off from the rest of the company and only authorized personnel are allowed into the area. Documents are thus processed by a clerk and equipment operators who specialize in this work. The advantages of this method are greater productivity, less labor wastage of more highly paid personnel, better performance and maintainability of equipment.

Of course, more than one reproduction facility may exist. In large companies, additional facilities may be placed near areas of high usage.

LOCATION

The central or main reproduction facility should be placed where it will be most cost effective. That is, it should be located nearest to the department that needs it most. Thus, walking time to and from the facility is reduced and notification or delivery of finished prints is made easier if the users are visible to the reproduction clerk. For example, if the time that designers must walk to and from the reproduction facility is reduced by only 30 seconds, the cost saving over a year for a 25-man design and drafting group will be $1200*, assuming that each designer makes two trips a day to the facility.

SIZE AND LAYOUT

The size of the facility, of course, depends on the staff and equipment needed. However, note that failure to allow for ready conversion of adjacent areas to reproduction use, can lead to expensive modifications or relocations. Because reproduction facilities require special layouts, partitioning, and power requirements, it is usually much less expensive to move out an adjoining group of engineers, designers, or clerk-typists than to relocate the reproduction facility. Thus, even if a stable and predictable capacity requirement is known, it is still safer to consider the ease with which adjoining groups can be moved to make room for a larger reproduction facility.

Although it is desirable to provide spare, unused growth space, its cost

*25 men x $\frac{2 \text{ x } .5 \text{ minutes x } 240 \text{ days}}{60 \text{ minutes}}$ x $12/hr (burdened) = $1200

may be prohibitive. For example, assigning 500 square feet of unused space to the reproduction facility will cost the company $4700 a year if lease or rental costs are 80¢ per square foot per month.

Guidelines for determining the size of the facility follow:

1. Number and size of reproduction machines.
2. Number of work tables and surrounding open space required.
3. Number of files required (regular, legal, and flat).
4. Number of storage cabinets for paper, cleaning supplies, and other materials.
5. Number of desks needed.
6. Number and size of additional equipment to be purchased during the next 2 or 3 years.
7. Number and size of accessory items: binding equipment, paper cutters, collators, folding machines, etc.
8. Minimum space requirements between reproduction machines and walls for cooling and repair.
9. Expected percentage staff growth rate per year.
10. Amount of storage area needed for keeping a minimum (1 or 2 weeks) supply of reproduction materials.*

The specific layout of equipment and furniture can vary, depending on the operating mode and constraints applied to the facility. However, several points must be kept in mind in the layout of the facility. A checklist is given in Table X.

Table X

Layout Requirements Checklist

1. Location of electrical power outlets of required voltage and current capacity.
2. Availability of separate power lines for reproduction machines, so that overloads in other parts of building will not affect the reproduction equipment.
3. Availability of a special area or access to outside for keeping high pressure ammonia containers for whiteprinters.
4. Location of ventilation ducts for ammonia fumes.
5. Allowance for adequate open space near reproduction equipment for operation and maintenance.
6. Convenient location of paper supply to minimize time required by operator to get paper for making prints.
7. Adequate size doorways for moving equipment in and out.
8. Sufficient excess of outlets for connecting new equipment.
9. Location of waste containers or paper shredders.

*This refers to immediate reproduction area. Additional materials for a longer period are stored elsewhere.

10. Location of area for additional desks and file cabinets.
11. Availability of a separate storage area for keeping unused ammonia containers and for discarding old amonnia.
12. Adequate space for on-line sorters or collators.
13. Locations of sprinklers and fire extinguishers.
14. Location and size of vault area.

EQUIPMENT, CABINETS, AND FURNITURE

A wide range of equipment and other items are required for an efficient reproduction facility. While the reproduction equipment is essential to the operation, accessory equipment and associated cabinets and furniture play a major role in the overall effectiveness of the reproduction facility. Key reproduction equipment includes:

- Whiteprinters/blueline machines
- Xerox or equivalent blackline machines
- Microfilm-to-hard copy machines

SPECIAL CONSIDERATIONS

Several important areas related to the reproduction facility and its equipment should be considered in establishing a new capability or improving an old one. These include:

VENTILATION—Adequate ventilation is needed to minimize the accumulation of white printer ammonia fumes. Ammonia can be fatal to personnel if not vented properly. Note, however, that some whiteprinters do not require venting.

HUMIDITY—High humidity is bad for machines, reproduction materials, and people. Therefore, room air should be kept comfortably dry.

TEMPERATURE—The temperature should be kept on the cool side because of the physical activity involved compared to sedentary desk work and the deteriorating effects of high temperatures on sensitized reproduction materials. Temperature fluctuations may also affect copy quality when whiteprinters are used.

POWER REQUIREMENTS—Adequate electrical power capacity and outlets must be provided for present and future needs. Some machines need their own power lines without have any other equipment run off them. In addition, 220 volt outlets are often needed.

FLOOR STRENGTH—Although this is normally not a problem, old office buildings may require checking to ensure an adequate safety factor when a number of large machines are to be located on a wooden floor.

NOISE—Most machines produce some noise when operating. Therefore, their location near areas where people require silence to concentrate on their work is not desirable. Otherwise, additional soundproofing must bc addcd to minimize noise transmission.

8.4 Reproduction Equipment

Many types and styles of reproduction equipment are available with various speed and document size capacities. However, the basic types commonly in use are the whiteprinter, xerographic and microfilm-to-hard copy machines. The basic characteristics of reproduction machines are given in Table XI. Descriptions of each machine type follow.

Table XI

Reproduction Equipment Characteristics

- Maximum printing width size (e.g., 11 to 54 inches)
- Exposure lamp type (e.g., mercury arc or fluorescent)
- Lamp wattage (e.g., 200 to 7500 watts)
- Maximum speed (e.g., 1 to 75 ft/min or 3600 copies/hr)
- Operating voltage (115V, 60 Hz, or 230V, 60 Hz single phase)
- Maximum operating current (e.g., 7 to 75 amperes)
- Method of separating original and image development (e.g., automatic/manual)
- Maximum dimensions (e.g., 15 by 20 by 32 to 70 by 90 by 80 inches)
- Weight (e.g., 50 to 2400 pounds)
- Type of feed (e.g., sheet or roll)
- Odor/ordorless
- Fume vents (required or not required)
- Exhaust diameter (e.g., 8 inches)
- Exhaust rate (e.g., 0 to 1000 CFM)
- Method of delivery (e.g., copies are ejected from rear, front, or side)
- Ammonia type (aqueous ammonia or anhydrous ammonia)
- Operator skills required
- Reliability (e.g., mean-time-between failures)
- Simplicity/ease of operation, maintenance and repair
- Inventory requirements (e.g., supplies, spare parts, etc.)
- Noise level
- Versatility (e.g., size of originals that can be used to make copies, ability to make copies from books or opaque materials)
- Automatic sorter attachments, including capability to increase maximum number of copies that can be sorted at one time
- Temperature and humidity limits (e.g., 60 to 90°F, 15 to 85% humidity)
- Warm-up time (e.g., none to ½ hour)

WHITEPRINTERS (BLUELINE, OZALID, AND DIAZO)

The whiteprinter is the traditional machine for copying engineering drawings and data. It can be obtained in several types, but the basic difference is in the size and speed selected. For example, a 50-pound bench machine is available for reproducing drawings up to 42 inches wide by any length at speeds up to 6 feet/minute. Its overall dimensions are 52 by 8 by 8½ inches. On the other side of the spectrum, a 2400 pound machine is available for handling up to 54 inch wide drawings at speeds of 75 feet/minute. This machine is 70 by 105 by 50 inches. Prices vary from several hundred to over $23,000 dollars.

Whiteprinters operate on the basis of light sensitive chemicals. That is, they require paper coated with a material that undergoes a chemical change when exposed to light. The change is such that a latent image of the original document is created on the copy paper. When this copy paper is exposed to ammonia, or other chemical, the latent image is converted into a positive image of the original.

Proper selection of a whiteprinter can save a company money.

> **CASE IN POINT: A medium size company saved $16,000 a year by installing a table model white printer in its drafting area. It eliminated having draftsmen, checkers, designers, and engineers from wasting their time waiting for checkprints from the central reproduction area. This service required completion of a requisition form and a 2 hour to 2 day wait to get a checkprint. Using the whiteprinter, open to all in the engineering department, anyone needing a checkprint could just walk over to the machine, turn it on, and make a print in minutes. No waiting was necessary even though the machine was used many times a day. The cost of the machine was small compared to the large savings in expensive technical manpower. In addition, design and drafting work was released sooner because there were no delays in getting checkprints.**

SPECIAL CONSIDERATIONS—Special considerations related to whiteprinters are primarily concerned with the use of ammonia and environmental conditions. Some machines release ammonia vapors into the air and must be vented or quickly removed from the area. In addition, ammonia must be periodically replaced with fresh material. For larger machines, high-pressure bottles are used. To ensure personnel safety, these bottles should be kept in a separate room and connected to the machine through special lines.

Since the whiteprinter depends on a chemical process to copy originals, temperature and humidity affect the speed and quality of the reproduction made. Therefore, stable temperature and humidity ensure more uniform and higher quality copies.

BLUEPRINTERS

A blueprint is a copy of a drawing with a blue background and white image areas or lines. Blueprint making is one of the oldest methods of reproducing drawings. Recently, however, it has been replaced by other techniques. The process is slow and requires washing and drying before a final print is obtained.

The blueprint process works this way:

- Paper coated with a special chemical is obtained.
- The original vellum is placed over the paper and exposed to light.
- The coating not covered by lines on the original is altered in properties so that when washed in water produces a blue color.
- The portion of the paper not exposed to light has the water-soluable chemical washed away, leaving white lines on blue paper.

This method of making prints is still used in architectural and structural work.

XEROGRAPHIC COPIERS

Because of the wide popularity and variety of xerographic copiers, this section is devoted to this type of reproduction equipment. Some representative types of xerographic copiers follow:

- Standard letter-size copier
- Combination standard copier and reducer
- Engineering data printer and reducer
- High-speed engineering data printer (also folds and sorts engineering prints)
- Microform to hard copy printer

An example of how the right machine can be used to save a company money follows:

> **CASE IN POINT: An opaque copy reproduction system was installed in a small company because of its greater efficiency over existing duplicating equipment. These savings were achieved because this xerographic machine could automati-**

cally produce reduced size, folded copies of vellums much more rapidly than by using a small blueline printer.

The copier could print, fold, and sort up to 50 prints from a D or smaller size vellum at a fraction of the cost of the blueline machine. The annual savings using this xerographic copier was over $4000 a year.

Xerographic printers vary widely in cost per month. Depending on copying needs, monthly rentals can range from $100 to $1000 and more.

PRINCIPLE OF OPERATION—Xerographic reproduction works on the basis of light neutralization of electrostatic charges. A drum or plate coated with selenium is covered with electrostatic charges before it is exposed to original. As it rotates, the image on the original is exposed to a bright light. The image lines on the original absorb the light while the white/clear areas reflect it. The reflected light is focused by a lens system onto the drum surface. As the charged drum rotates, the areas that receive reflected light are neutralized forming a latent (invisible to human eye) electrostatic image of the original document on the drum. Next, an oppositely charged black powder is automatically sprinkled evenly over the drum. It adheres to the drum areas where charges remain (image areas or lines on original) but does not stick to neutralized or no-charge areas. The drum then comes into contact with a sheet of paper and transfers the powder image onto the paper. This transfer is achieved by a combination of pressure and electrostatic forces. The paper passes under a fuzer element which bakes the powder onto the paper for a permanent bond.

SPECIAL CONSIDERATIONS—Xerographic copiers have certain characteristics to keep in mind during their selection and evaluation for use in the reproduction facility or other areas in the company.

- The rental or lease basis for obtaining these machines allows easier upgrading of the reproduction facility because a newly developed copier can be exchanged for the old one without the need to sell the old one or to acquire new capital for purchase.
- Xerographic copiers are usually more expensive than other types of copiers. However, service and maintenance are free.
- Xerographic copiers are faster when many copies must be printed from a single original. (For over 20 copies, offset may be cheaper.)
- Glass plate for holding original can be scratched creating black marks on copies and drum, which must be cleaned periodically, can also be scratched causing poor copy quality.

8.5 Reproduction Media

A large variety of reproduction materials is available for the various types of reproduction equipment used in industry. Some materials are plain white bond paper, others are specially treated synthetic products such as polyester. The particular materials best suited to a company depend on its needs and budget. Some factors to be considered in the selection of reproduction media or materials are:

- Quality of reproduction
- Transparency or speed of reproduction (the more transparent, the faster the copying process when a whiteprinter is used)
- Durability of image copied and the medium itself
- Fade resistance
- Tendency to become brittle with age
- Dimensional stability
- Erasability
- Writeability (ease with which pencil or ink changes can be made on the medium)
- Absence of ghost images when corrections are made on an intermediate vellum
- Water proofness or sensitivity to moisture
- Tendency to discolor with time
- Loss of translucency with time if a vellum
- Resistance to the effects of heat and humidity
- Cost per copy
- Percent of waste or spoilage

A summary of the key types of reproduction media commonly used is presented next. Note that many varieties exist within each basic type and from manufacturer to manufacturer. For detailed information obtain catalogs from suppliers.

* Blueline paper (blue image on white paper base)
* Blackline paper (similar to blueline paper but has black images on white paper)
* Colored sensitized paper (blue or black image on pink, blue, yellow or green base paper)
* Plastic surface sensitized paper (sensitized paper for whiteprinter. Used where greater durability is needed and where dust and dirt are problems)
* Opaque sensitized cloth (durable sensitized medium for whiteprinter; durable but can be replaced by superior polyester medium)
* Sensitized cardstock (white printer; used when self supporting prints are needed for presentations or vertical filing)

* Unsensitized bond paper (used for xerographic reproduction)
* Overhead projection film (sensitized for whiteprinter reproduction; used with overhead projector which requires image area on clear base. Image can be in various colors)

REPRODUCIBLE/INTERMEDIATE

Whiteprinters and xerographic copiers can also be used to create images on translucent or semi-translucent media for subsequent reproduction on opaque copies. These translucent copies offer several advantages:

- A valuable original can be stored in a safe place while a translucent copy is used to make blueline prints for production and other uses.
- A permanent record of the original configuration can be retained even when changes have been made to the original image. (This is a history reproducible described in Chapter 3.)
- Several identical reproducibles can be made and used simultaneously to increase production when many copies are needed fast.
- Redrawing time can be saved when minor modifications or different versions of the original are needed. For example, several reproducibles can be made, modified, and renumbered to identify each configuration without redrawing the entire configuration. (Lines on the reproducibles can be changed with chemical eradicators or a wet eraser, depending on the type of medium used.)

Reproducibles come in three basic media: paper, cloth, and film (e.g., acetate and polyester). The type selected depends on its use, cost, and personal preferences. The advantages and disadvantages of each type are given in Table XII. Note that reproducible media come in various sizes, and image lines are either sepia or black. For sensitized media, shelf life may be fairly short; e.g., one manufacturer indicates that unexposed film will give good copies for about 4 months after shipment. (Of course, the developed medium may give good copies for an indefinite period.)

Table XII

Comparison of Different Reproduction Media

Medium	Advantages	Disadvantages
Paper	Lowest cost Can be folded and stored in regular files	Least durable Easily torn and dogeared Yellows with age

	Lighter weight	Becomes brittles with age Water sensitive Loses translucency with time Requires chemical eradicator
Cloth	More durable and moisture resistant, More workable surface, & Stronger than paper	Can be torn Not waterproof Most expensive
Film*	Dimensional stability Almost indestructable Faster reproduction Insensitive to heat Waterproof Permanent image Image lines can be erased without chemicals (some film types only) Accepts pencil and ink (some types only)	Expensive compared to paper More waste when used by inexperienced operator

Paper and cloth reproducibles are generally duplicated in the same way as regular, opaque sensitized paper. That is, the original is placed on top of the sensitized paper and run through the exposure chamber together. Then the exposed paper is passed, sensitized surface up, through the developer. Films, however, are handled somewhat differently. The original is often placed over the sensitized side of film (glossy side if one side has a matte finish for writing or inking) so that the image lines make direct contact with the sensitized surface. The resulting image is sharper and clearer than if it were done in the conventional manner. Copies are made by placing the image lines on the film in direct contact with the sensitized surface of the reproduction medium (paper, cloth, or film).

HANDLING AND CARE OF REPRODUCTION MEDIA

The handling and care of sensitized reproduction media are important items in the reproduction facility. Improper care can result in higher operating costs because paper, cloth, and film have to be discarded because of (1) premature deterioration of the chemical coating or (2) the media are folded, torn, or wrinkled. Therefore, to avoid these added expenses, plus the lost time of having to wait for another shipment if the media cannot be used at all, these rules should be followed:

1. Leave sensitized media in their protective bags or wrappings, unopened until needed.

*Acetate film is cheaper but less durable then polyester film.

2. Reseal bags or packages if a small amount of the media are used and the remainder will not be consumed in the immediate future.
3. Keep all media not needed immediately in dark, cool, and dry storage areas. (One manufacturer recommends keeping temperature between 62 and 65 degrees Farenheit at a humidity of 50 percent.)
4. Keep media away from ammonia fumes whenever feasible.
5. Use sensitized media on a first in, first out basis because they have finite shelf lives; e.g., 6 months.
6. Ask supplier for advice on handling and storage of media.

SPECIAL CONSIDERATIONS

As was mentioned before, reproduction media come in a great variety of characteristics and materials. When considering selection of the best media for your particular facility, ask the supplier for specifications on the following characteristics and his recommendations related to the needs and budget of the facility:

1. Reproduction speed (How fast can the paper be run through the machine and still obtain good quality prints?)
2. Deterioration of coating (How long can the medium be stored before use without serious degradation of print qualities?)
3. Deterioration of image on copy (When will it begin to fade? When will the paper begin to yellow and become brittle? Note that white prints will fade when exposed to sunlight; if filed properly, prints can last for about 5 years.)
4. Strength and durability (How easily can the medium be damaged? How much use can it see before it becomes worn out?)
5. Dimensional Stability (Will the medium's physical size change with temperature and humidity?)
6. Moisture and heat sensitivity (Will moisture and heat affect the medium? How much? What are the best temperature and humidity levels for storage?)

Note that technological developments are constantly producing new reproduction media and improving existing media. Therefore, periodic review of manufacturer's catalog may result in discovery of a new or improved medium that reduces reproduction labor or waste or offers new applications.

8.6 Staff

The staff of the reproduction facility does not need a high level of education or skills to effectively perform the operations needed to copy, store, and distribute enginering documents. However, proficiency is needed in the following areas:

1. Filing documents and document control procedures
2. Equipment operation and maintenance (repair skills not needed)
3. Record keeping
4. Typing
5. Care and handling of sensitized materials and original documents.

While extensive or in-depth skills are not needed, certain personal characteristics are critically needed for a good operation. These include consistent following of standard procedures, attention to details, and ability to increase output sharply for short periods in response to urgent requirements. Amiability is also a desirable quality. The pressures of the job and the natural tendency of requesters to want their documents copied before anyone else, can sour the staff's disposition and affect productivity and relations with the rest of the company. Therefore, short-tempered, moody personnel are not suitable unless reproduction requirements are modest and conflicting demands by the requesters are rare.

The size of the staff, of course, depends on the reproduction needs of the company. Normally, a minimum staff includes a clerk for receiving and feeding copy requests to the reproduction operator. If occasional periods of high copying requirements exist, personnel in other departments such as typists and clerks, can be trained to operate the reproduction equipment as emergency helpers.

Some training is required unless experienced personnel are available. Specific training areas include:

- Completion and filing of reproduction request forms
- Operation and maintenance of reproduction equipment
- Characteristics and handling of reproduction media
- Copying priority system
- Filing and distribution of prints and originals
- Stamping of prints with correct usage information and restrictions (see Chapter 3)

8.7 Cost of Reproduction

The major cost elements involved in the reproduction of engineering documents are the equipment used, the materials and supplies needed, accessory equipment, and labor required.

Copy costs are often measured in terms of cost per standard size sheet or per square foot. A comparison of the cost per standard size sheet for xerographic and whiteprinter methods is given next. (Costs include equipment and materials but exclude labor.)

Method	Cost per Sheet
Xerographic	4¢
Whiteprinter	2¢

While xerographic costs appear higher, convenience and lower labor costs also affect selection of the best method. For example, the total cost to reproduce 25 copies of B size drawing on xerographic machine can be less than the cost to copy them on a table model whiteprinter because they can be made in less time.

COST SAVINGS TECHNIQUES

Several techniques are available for reducing copying costs. These include use of satellite copiers, reduced walking time, use of an expediter, and others. Each is described next.

SATELLITE COPIERS—By placing low-cost satellite copiers near people who have frequent need for copies, the users can obtain their own copies immediately and with a shorter walk. For example, a copier can be placed in a D&D room where each draftsman can see the copier and obtain a copy when the machine is not being used. Thus, waiting time is also eliminated or greatly reduced.

PLACE FACILITY CLOSEST TO BIGGEST USERS—Considerable labor costs of expensive people can be incurred simply from the time lost in walking to and from the reproduction facility. Therefore, the closer the facility is to the largest and most frequent users, the cheaper it will be. A $1200 annual cost was computed in Section 8.3 for a 25-man design and drafting group. This was a conservative estimate because it did not include the labor lost when the draftman stops to talk with someone on his way to the reproduction area. It is safe to assume that the longer the walk, the greater the chances of the draftsman stopping somewhere along the way.

SPECIAL MESSENGER/EXPEDITER—When the demand warrants it, a low paid messenger can be used to pick up documents for reproduction from the draftsmen, designers, or engineers and to bring them to the reproduction facility. He can then pick up the prints when they are ready and distribute them. For a large facility, the messenger can have his station located so as to be visible to the users who can then call him to their work areas to pick up an original for copying.

REDUCING REPRODUCTION MEDIA COSTS—Large orders of a few types of reproduction media can be obtained with larger discounts and lower costs. Also lower purchasing costs will result due to fewer purchase orders being prepared. Therefore, the ability to accurately predict the amount and types of reproduction media needed can reduce prices paid and waste due to lack of use and deterioration of purchased media. Note that a special storage facility kept at a constant and cool temperature can reduce wastage by extending the reproduction media's shelf life.

INCREASE PRODUCTIVITY OR REDUCE COST PER COPY—By using fast* papers, the labor required to make prints is reduced and more copies can be run through the machine, avoiding the need for another machine. The cost per copy can also be reduced by locating materials and accessory equipment as close to the operator as possible. This minimizes time lost due to walking to and from where the sensitized paper is stored. The preceding applies to whiteprinters. When xerographic copiers with large capacity feeding trays are used, the location of storage areas is not as critical. However, spare paper should still be quickly available to the operator.

NIGHT SHIFT—A night shift or extended day allows reproduction facilities to be used longer, resulting in a greater monthly output capacity. This increased output capability can avoid the need for purchasing or renting more machines and hiring more prople. Also the need for a larger facility is avoided. Copying machine rental companies offer reduced rates when large numbers of copies are produced each month. Different plans are available. The sales representative should be contacted for finding the best one for the company based on current monthly copying requirements.

REDUCING MAILING COSTS—If a company mails many drawings each month to customers, field engineers, and others, it can obtain substantial savings by using reduced size copies. For example, D and C size originals can be reduced to B or A size copies with a subsequent reduction in mailing costs. Microform can also be used to reduce mailing costs if the savings justifies the expense of microform equipment for the company and users of the microforms.

> **CASE IN POINT: Use of microfiche can reduce mailing costs considerably. For example, mailing a 50 page specification on microfiche costs only the regular first class rate. Mailing this same document in normal print size costs about $1.00. For companies that mail drawings, specifications, catalogs, and reports out extensively, the use of microfiche can save thousands of dollars a year in mailing costs. In addition, mailing envelopes can be smaller and cheaper.**

REDUCING FOLDING COSTS—Folding drawings can also be expensive and can limit the output capacity of the duplicating machine when the people folding can't keep up with the people making the copies. Thus, the use of an

*At the usual cost of lower contrast image lines.

automatic folding machine can lower labor costs and can more than pay for itself when much folding is required. The use of a reducing machine can also save labor costs by allowing folding clerks to work with B size copies rather than C, D, or E size copies.

REDUCING STORAGE COSTS—Minimum printing of file copies can reduce the need for file cabinets and associated floor space. At 75¢ per square foot per month, this can lower facility costs by a large amount when many drawings must be kept by a company. Use of reduced size copies and microfilm also saves space.

ELIMINATING REPRODUCIBLES—the need for reproducibles (brownlines) as historical records (see Chapter 3) can be eliminated when opaque copies of large size drawings can be reproduced by methods other than whiteprinters. At a total cost of 50¢ each, eliminating 2000 reproducibles per year will save the company $1000.

> **CASE IN POINT: One company saved over $5000 in brownline paper and additional labor costs by eliminating the need for brownline intermediates.**
>
> **The brownlines, which cost more for material and labor to produce, were replaced with ordinary copies on white bond. Each copy was stamped "FILE MASTER." This copy was a historical document in case an engineer needed a reference print. This master file copy was used to produce a xerographic print when the engineer needed his own copy.**

REDUCING PERSONAL USE OF MACHINES—When copiers are open to anyone, they have a tendency to be used for personal use. Depending on the type of company and staff, non-business related copying can become expensive. It not only increases reproduction and repair costs, but wastes the user's time. The use of restricted reproduction areas reduces this cost automatically. However, open access machines are the best solution for small companies in many instances. Therefore, the personal use of copiers has to be discouraged by policy statements and memos.

8.8 Avoiding Problem Areas

Problems in the reproduction facility can affect the overall operation of the company and especially the engineering, purchasing, and manufacturing departments. If prints aren't available when needed for engineering review and approval, valuable engineering time will be lost and production delays will occur until the documents have been checked, corrected, and released. As a result, project costs will increase and scheduled deliveries will slip. Although slower than optimum turn-around on prints may not show any serious cost additions or delays, the cumulative effect of slow reproduction of originals on a company-wide basis can result in the loss of thousands of dollars per year. Therefore, attention to avoiding copying problems and their inevitable effects on efficiency or productivity is a cost effective operation. Some common problems are discussed next.

UNDERSTAFFING

Understaffing can create serious problems in the reproduction of copies at peak company demands for documentation. The effects, mentioned before, are delays, wasted staff time, and slipped schedules. The question here is: Is the money saved by understaffing the facility worth the losses incurred by the rest of the company? Of course, while the reproduction facility may be efficient appearing at the end of the year, its indirect effects on the other departments may have made it a serious liability. Therefore, before costs are cut in the reproduction area by keeping the staff small or reducing personnel, examine the effects of increased reproduction turn-around time.

CASE IN POINT: Understaffing in the reproduction area can be expensive. While a small staff will reduce the operating costs of the operation, it can increase overall company costs by much more than it saves. One company cut its manpower to a minimum to reduce costs. However, as time passed, filling of drawing requests slowed and finally deteriorated to a point where it was frequently taking 2 weeks for a designer or engineer to get a print of a drawing. The costs to the company in delaying completion of new designs and revisions increased cost in getting the products to the customer. For this company, a loss of well over $20,000 per year occurred.

One way to keep the staff small and yet have the potential for greater output during peak work loads during the year, is to train messengers, typists, and clerks to be able to step into the facility on special call to help the regular personnel.

Another approach is to make arrangements with a labor contracting agency to provide reproduction personnel within a day's notice. There are a number of agencies in large cities that can provide this service.

If the company is located in an industrial park, agreements with other reproduction facilities to help in emergencies may be feasible. This can be done with a previsously agreed to charge or credit rate per copy.

Use of overtime or a second shift can also help compensate for a smaller staff. The premium rates paid will usually be much less expensive than the cost of an additional machine and space for it. Remember that additional equipment and space increase utilities and maintenance costs on a permanent basis.

Of course, each of the preceding techniques has disadvantages such as lower total efficiency, more errors, lost originals, greater waste, and more coordination labor. These disadvantages must be weighed against the real or imagined cost savings in maintaining a staff that is too small to handle above average printing demands.

EXCESSIVE DELAYS AND WAITING TIME

Excessive delays and waiting time not only increase costs in other departments but generate bad feelings in non-reproduction personnel that can affect the morale of the reproduction staff. Therefore, it is important that both staff and users feel satisfied about the operation.

The solution to avoiding excessive delays for prints/vellums was mentioned before under Section 8.7 "Costs of Reproduction." However, it is important to note that action should be taken well before a critical condition in terms of excessive delays and waiting time occurs. Therefore, periodic checks should be made to observe when gradual but constantly increasing delays and waiting periods for finished prints require increased capacity.

POOR QUALITY COPIES

Poor quality copies can be created by inexperienced personnel, deteriorated reproduction media, excessive work loads, or poorly maintained equipment. To avoid this situation, periodic training, checking of the handling and treatment of materials and supplies, and procedures for cleaning and maintaining equipment are needed. However, sporadic inspection or surveillance systems will not ensure the avoidance of poor prints. They must be employed consistently.

Other factors that will affect the copy quality of whiteprinters are temperature and humidity. Variations in ambient temperature require periodic speed adjustments to ensure top quality prints. High humidity will also affect print quality by curling the paper. When using polyester film and other media, heat and moisture conditions in the developing chamber can cause the media to stick. Installation of a Teflon slip screen can help overcome this problem. The slip screen reduces friction and enables the sensitized media to be carried over the tank or rollers without sticking or creasing.

ERRATIC OUTPUT OR PERFORMANCE

Erratic output can often result from morale problems such as feelings of being unappreciated. The reproduction staff is often looked on as a necessary evil and most company personnel don't appreciate the problems and hard work that the staff has to endure. The projection of this lack of appreciation can often undermine the staff's desire to provide top quality service.

Another cause for poor performance is the lack of supplies and materials when needed. If the operator has to wait an hour or two for a rush shipment, his output is going to fall drastically. This is especially true if he has to run around to find substitute materials or scraps of blueline paper at different locations.

An old or poorly maintained reproduction machine, which breaks down frequently, is another source of poor performance. If the staff is constantly calling the service man or having to make adjustments to the machine, the output of prints is going to be reduced. Therefore, machines should be replaced or rebuilt before they begin to have frequent troubles.

FILL-IN WORK

Because print or copying demands are not always uniform, periods will occur where no copying is needed. This problem of temporary inactivitiy can be overcome by some planning before it occurs. The following measures can be taken:

1. Collect a backlog of low priority copying for slow periods.
2. Schedule training and review sessions with staff on equipment maintenance, new reproduction techniques, and materials handling.
3. Clean up facility and equipment more thoroughly than usual.
4. Take initiative in contacting various departments for work.
5. Search for ways to improve performance.

Examples of low priority printing include extra copies of standard forms, handbooks, technical articles, and product specifications. Draftsmen, designers, and engineers often need personal copies of an article or product specification that someone else has. However, they may not bother getting copies made because of the effort or time involved in going to the reproduction facility and filling out a reproduction request. Instead they will waste time asking someone else for his copy. When the time is available, the reproduction operator can ask the staff for documents that would be of interest to others, copy them, and distribute them.

9.0 Drafting and Design Costs

Drafting and design costs represent a major segment of industrial expenses. For example, about 12 billion dollars a year are spent preparing and revising engineering drawings. Such a large expenditure would indicate that reliable and consistent guidelines for accurate cost estimating are well established in industry and government. Unfortunately, in most cases design and drafting costs are rarely predicted with good accuracy, i.e., less than 10 percent error. There are several reasons for this situation such as inability to accurately predict (1) the exact number of drawings that will be prepared during a program; (2) the complexity of the drawings that will be needed to define the product; (3) the number and extent of revisions to drawings during the life cycle of the product; and (4) future labor costs and efficiency (changes in people, salaries, management, and procedures occur frequently creating an unstable base for estimating costs).

Based on the above, a pessimistic viewpoint may be taken about producing accurate D&D cost estimates. However, if one is aware of the pitfalls in estimating D&D costs and attacks the problem systematically, he will achieve better results than average. The following sections present some representative techniques for estimating D&D costs for a program or product development effort.

BENEFITS AND APPLICABILITY OF THIS CHAPTER

This chapter describes the overall operation of estimating drafting and design costs—an activity that is important to all engineering organizations. Specific benefits are:

—Provides various techniques for estimating costs.
—Gives a systems approach to estimating costs that covers the life-cycle of the drawing.
—Provides sample D&D costs by function and type of document.
—Presents special factors for modifying basic estimates.
—Gives a checklist for ensuring that the estimate is complete.

Systematic and formal estimating of D&D costs is necessary regardless of company size. Any company that assumes D&D costs will take care of themselves is going to find itself in trouble.

CASE IN POINT: One small company estimated D&D costs for design and development of a new product as a blanket 20% of the total engineering budget. No detailed examination was made of drawing requirements, risk of redrawing due to engineering and design changes, and manpower requirements. The result: the design and drafting effort turned out to be the most expensive area next to the cost of material and parts for the equipment. The project overran its D&D budget by over 200%

9.1 Cost Estimating Techniques For Reducing Overrun Risks

Several methods are used to estimate design and drafting costs. Some common ones are described next. Note, however, that the method used depends on the estimator's personal preferences, company policy, and experience. It is recommended that the simplest and most straightforward method that proves effective be followed to minimize errors and laborious calculations.

HOURS PER SQUARE FOOT

One popular method for estimating D&D costs is to determine the hours it takes on the average to prepare 1 square foot of a drawing. This cost factor may be obtained from published information or by calculating the total hours spent by design and drafting personnel in preparing drawings over some period such as the previous year. Thus, if a company produced 1000 D-size (5.2 square feet x 1000 = 5200 square feet) and associated revisions over the last year and paid its people for 27,040 hours of work, then the hours per square foot are 5.2. Assuming the nature of the drawings to be prepared for the new year is the same, the cost to prepare 2000 D-size drawings can be easily and accurately computed.

If an average size drawing isn't used, then the number of square feet is obtained by multiplying the quantity of drawings for each size by their respective square feet such as:

Drawing Size	Square Feet
A	.65
B	1.3
C	2.6
D	5.2
E	10.4

A-UNITS

Another system for estimating costs is to use what is called the A-unit. An A-unit is equivalent to an A size drawing (8½ by 11 inches). Each successive increase in drawing sheet size results in double the area of the drawing. Thus, a B size sheet is twice the area of an A size sheet. Likewise, a C size sheet is twice the area of a B size. Therefore, if we know the hours needed to prepare an A size drawing (.65 square foot), we merely have to double the hours for each next larger size to get the hours per drawing. Once the hours for a basic A drawing have been determined, the following list can be used to convert each drawing size into an equivalent preparation time:

Drawing	Equivalent A Units
A	1
B	2
C	4
D	8
E	16

How much is the Basic A unit worth? This depends on the type of business, checking requirements, type of program or product, economic conditions, and the specific company's efficiency. Estimates range from 2 to 8 hours per A unit.

TYPES OF DRAWINGS

Another approach is to determine the types and complexity of the drawings that will be needed for the program. Predetermined hour assignments are used to calculate the total drawing effort cost. Thus, the number of assemblies, detail parts, printed circuit artworks, list of materials, castings, chassis, covers, valve and wire lists, schematics, logic and piping diagrams are estimated. These drawings are then assigned complexity factors, depending on the product being developed, and multiplied by the appropriate hours. The sum of all the hours needed for each drawing gives the total hours to complete the effort.

Some sample costs by drawing type follow:

Drawing Type	Hours to prepare
Schematic	24
Design layout	40
Tape up (master artwork)	20
Detail fabrication board	8
Board Assembly	10
Parts list	3
Revisions	24

DRAWING SIZE AND COMPLEXITY FACTOR

Drawing size and complexity factor are also used to estimate drawing costs. This system may ignore the types of drawings being prepared and concentrates on paper size and the details needed to depict the part or assembly. For each drawing size, different complexity factors are used such as 2.2, 5.3, 8.3, and 11.4. These factors are multiplied by the area equivalent to the drawing size needed and converted to total hours per drawing. The sum of the hours for all the drawings gives the hours of labor needed to prepare the drawings needed for the program or product.

9.2 Basic Approach to Estimating

The basic approach to estimating design and drafting costs is described next. This method involves consideration of the life cycle cost of a drawing, historical cost data, estimating the number of drawings needed, and calculation of the total cost in hours.

LIFE CYCLE COSTS

Preparation of the intial drawing is usually the largest part of the drawing life cycle cost. However, revision and maintenance costs during production and sustaining phases represent a large part of the total cost involved in preparing and using a drawing. Therefore, if an estimate is needed for a program that will involve drawing and design costs beyond the initial drawing preparation stage, additional costs must be added to the basic cost estimate. Note that these additional costs may equal or exceed the initial D&D estimate.

CASE IN POINT: One company failed to estimate drawing life cycle costs for a new product development program. As a result, it estimated the average cost of a drawing at 25 hours.

However, because of many design changes and value engineering improvement for lower production costs, the average cost per drawing was found to be closer to 70 hours by the time the program was over.

Other cost elements that need to be considered in the life cycle costs are checking, administrative, reproduction, microfilming, and clerical labor. Although engineering also spends time checking and reviewing drawings, this cost is usually included in a separate estimate prepared by engineering.

HISTORICAL DATA

As mentioned before, historical data for a particular D&D department is needed to accurately assess the cost of drawings for a new program. Externally obtained cost data should be used only when no other reliable data are available within the company. However, care should be taken when using historical cost data to be sure of the following:

1. Cost data include all life cycle costs, including design, checking, administration, and re-drafting and all elements are defined; e.g., if the cost per drawing is given, how many sheets does this include or does it really mean cost per sheet? How many sheets per drawing were used?
2. Cost data reflect true cost of drafting effort. If the company plays games with charge numbers, then engineering costs may appear as drafting costs or vice versa. As a result cost data are not reliable for estimating excepting as a rough guide.
3. Is the new program similar to the ones conducted in past? If a company has only done commercial work and then is required to work for the government to tight specifications and drafting standards, costs could skyrocket by 50 to 100 percent.

If the company does not have reliable historical cost data, then a program should be conducted to obtain it from old records or to begin a systematic effort to collect reliable data on new efforts. Note that a secondary benefit of accurate cost data from old and recent efforts is that it allows a more objective evaluation of trends in efficiency and productivity.

ESTIMATING NUMBER OF DRAWINGS NEEDED

Accurate cost estimates on the hours per square foot are useless unless an accurate base is available on the toal number of drawings to be produced for the product. Determination of drawings depends on the cooperation of engineers, designers, and draftmen. The D&D department cannot determine an accurate number of drawings needed without consulting with engineering to determine whether old drawings can be used, the complexity of the product, and various types of miscellaneous drawings required.

A product breakdown chart or drawing tree from engineering can help considerably. However, D&D should not follow this input blindly because engineers often ignore minor detail drawings, which can add up to considerable hours because even a simple drawing needs to go through the entire process of number and nomenclature assignment and control, format and quality checks, and administrative processing and controls. Drawings and documents frequently overlooked, short-changed, or unrealistically estimated are:

Nameplates
PC board extenders
Miscellaneous, small parts
Covers
Cases
Gages
Test and assembly fixtures
Specification control drawings (may be 10 to 20 percent of all drawings produced for a program)
Separate parts lists & data lists
Standard parts drawings
Cables

COMPUTING ESTIMATE

Once the total drawing number is obtained, it can be converted to square feet* by using the various drawing sizes or by using a standard drawing size such as a D, to obtain total square feet of drawings to be produced. This number is then multiplied by the standard hours per square foot. Depending on the company, program manager, engineers, and type of work, a risk factor may be added to cover unexpected increases in cost due to higher than normal revisions, inaccurate input data, and drafting schedule compression due to delays in inputting concepts or designs to D&D. A 10 to 20 percent safety factor may be added to the total hours computed for the effort, depending in the uncertainty of drawing requirements.

Once the hours have been computed, these can be converted to dollars by using the average burdened hourly rate of D&D and multiplying it by the total hours. This quantity times the General & Administrative (G&A) rate is added to the previous product. This number yields the true cost of the effort.

*or to other units by using one of the methods described in Section 9.1.

Thus,

$$\text{Total D\&D cost} = \text{total hours x burdened rate} + \text{G\&A rate x (total hours x burdened rate)}$$

Depending on the company, the burden rate may cover materials and outside services needed for drafting. However, in other companies the costs of materials and special outside services may have to be added to the above cost. Note that material and service costs are increased by the G&A rate times the total material and service costs.

MODIFICATION OF BASIC COST ESTIMATE

The basic cost estimate should be modified up or down when conditions differ from past programs. These conditions are described in Section 9.4.

9.3 Sample Costs

It is impossible to give costs that apply to every company, program, and drafting activity. However, some sample costs in terms of hour figures are given in this section as a baseline when historical data are not available for making an estimate. Note that these figures do not represent a particular company's D&D costs and that under specific programs, costs may differ widely above and below these figures.

DESIGN

Design labor often lies between 20 and 40 percent of drafting costs. Thus, if a drawing will take 30 hours to prepare, design time will range between 6 and 12 hours. The higher percentage usually applies to the R&D phase of a program while the lower one is more applicable to early production operations requiring minor design changes. Note that some drawings such as construction detail sheets, may need more design time than drafting time.

DRAFTING

Drafting costs can range from 2 to 12 hours per square foot of drawing produced. Thus the labor for a D size drawing could be 10.4 to 62 hours over its life cycle. However, the initial (R&D) preparation cost alone should range from 8 to 35 hours. Supervision and administrative labor is added to the above hours per square foot estimates.

CHECKING

Checking may run between 20 and 50 percent of drafting labor. Sometimes it may be as low as zero or as high as 100 per cent of drafting time. At 25 percent of drafting hours, checking labor for a 50 hour D-size drawing would be about 12 hours.

SUPERVISION AND ADMINISTRATIVE

Supervision and administrative costs were omitted from the estimates given for design, drafting, and checking. Therefore, additional hours have to be added to the preceding numbers for a complete estimate. Hours for these functions range from 10 to 20 percent of all hours charged to the drawing. If supervision and administrative labor are indirect and part of the overhead budget, then these hours can be omitted from the estimate.

ENGINEERING ORDER/CHANGE NOTICES

The labor for these documents may be included in the drafting and design estimates described before. If they are not, then a separate cost computation is needed to cover these additional hours. When the drafting department is responsible for preparing all EO's/CN's, the hourly breakdown by job category can be as follows: Design: 3 hr, Drafting: 5 hr, and Checking: 2 hr.

CASE IN POINT: EO/CN preparation can be extremely expensive. One medium-size company prepared 5000 EO/CN's in one year. The combined D&D labor costs were over $½ million. Many of the changes were to correct design and engineering errors due to pressure to get work done on short schedules or inadequate design reviews.

If experience shows that an average of three EO's are prepared per drawing, then the total hours for EO's when 300 drawings are needed are:

EO hours = 300 dwgs X 10 hours/EO X 3 EO's/dwg = 9000 hours. It is obvious from the above that design/engineering changes add up to major program costs and are often the reason for cost overruns.

SEPARATE PARTS LISTS

When separate parts lists are used these may be estimated as a part of each drawing or on a separate basis. Parts lists require design, drafting, checking, and supervisory labor just as drawings do. Labor costs may vary from 1 to 10 hours per list, depending on customer requirements and the number of sheets needed. If the estimate includes changes during the parts list's life cycle, then no extra hours need be added. However, if the estimate, for example 3 hours per list, does not include revisions and changes, then the average number of changes per list must be determined and multiplied by the average number of hours needed to implement an EO or CN.

SPECIFICATION CONTROL DRAWINGS

Specification control drawings require about 12 hours of design, drafting, checking, and supervision time. The assistance of a Standards engineer is also needed and his time appears in another cost center or area. If considerable textural material is needed, a typist will also be included in the labor hours for these drawings.

TEST SPECIFICATIONS AND PROCEDURES

Engineering usually prepares test specifications and procedures, but drafting and checking labor is also required. These documents are multisheet, A size and require about 10 hours of drafting and checking time for an average 5 to 10 page document. In many companies, test specifications and procedures are prepared by documentation engineers and the publications department. If this is the case, no D&D costs need be included for this work.

REPRODUCTION

Duplication costs run 4¢ to 15¢ per square foot, including labor. At 6¢ a square foot, 10 copies of a D-size drawing cost about $3. Brownline and Mylar* reproducibles may be required by the customer. Brownlines are more expensive to print than whiteprints because the paper costs more and the duplication process is slower (more labor). Mylar, however, is considerably more expensive, e.g., a D size sheet costs about $12. If 500 sheets have to be duplicated on Mylar, the total cost will be $6000.

Note that in some companies duplication may be charged to an overhead number and need not be included in the D&D estimate.

DATA SUBMITTAL

A data submittal is a formal submission of drawings and associated lists and test specifications to the customer. If microfilm is required as a part of the submittal, the full-size documents must be inspected for microfilmability. When needed, lines and lettering are made darker or sections of the drawing are redone to improve quality. After microfilming, the reduced documents are reinspected for acceptable quality. The cost of this task ranges from 1 to 3 hours per drawing, excluding microfilming and computer card preparation. These costs are extra.

CLERICAL

Clerical labor is often a part of the overhead budget and may not be

*Registered trade mark of DuPont.

charged to the D&D effort directly. However, if typing and filing work directly related to the D&D effort is charged to the program or effort, the D&D estimate must include money to cover this work. Clerical labor should be between 5 and 10 percent of the total drafting labor, depending on the amount of typing to be done.

REVIEW AND SIGNOFF

Review and signoff operations are primarily engineering functions. The time to review, reject or ask a question, and sign off a drawing can vary considerably depending on the complexity and size of drawing. An average of 2 hours per drawing over its life cycle is a reasonable figure. Since this cost is part of the engineering, manufacturing, and quality assurance functions, it should be omitted from the drawing cost estimate. Labor for the designer to sign off a drawing is included in the estimate given under *Design*.

OVERHEAD AND GENERAL AND ADMINISTRATIVE RATES

Overhead (burden) and General and Administrative (G&A) rates vary considerably from company to company. Overhead rates include rent, taxes, management salaries, materials, and maintenance costs. Note that the cost elements included in overhead and G&A rates vary from company to company. Overhead and G&A rates are available from accounting when needed. However, for medium to large companies, the D&D department submits raw cost estimates which are converted by the accounting or pricing department into fully burdened costs.

Overhead rates may vary from 30 to 200 percent while G&A rates vary from 15 to 30 percent. When these rates are applied to a raw estimate, the full cost per hour of D&D may range from $10 to $16 per hour.

SUMMING UP

An example of design, drafting, checking, and supervision labor hours is given next for a D size drawing over its life cycle. (Drawing changes due to retrofit of delivered equipment are not included.)

Function	D&D to Release	Sustaining (post release changes)	Total
Design	13	4	17
Drafting	28	9	37
Checking	9	2	11
Supervision	5	1	6
	55 hours	16 hours	71 hours

The preceding hours were used by a company for preparing drawings for complex electronic equipment to high drafting standards and tight controls. Small companies or large companies producing simple items will find their hourly cost per D size drawing much less. Therefore, the above table should be used as a rough guide and adjusted to the particular operation.

9.4 Adjustments to Basic Estimates

Once the basic estimate has been formulated, it should be reviewed for variation in the D&D effort from the work done in past years. Most factors will probably be the same; however, a few may deviate considerably from previous programs and could make a significant difference in the estimate. Therefore, adjustments, either up or down, may be needed. Consider the following items before submitting the final estimate to management.

SIMILARITY TO OTHER JOBS

If the overall D&D effort is similar to that used for historical cost data, no change is needed here. If, however, the new project will be considerably different, a percentage difference should be applied to the estimate to account for a higher or lower total cost. Thus, if engineering states that this product is 10 percent more complex than previous ones, the D&D costs should be increased by 10 percent or more. Why more? Because complexity and hours per square foot/drawing may not be linearly connected. That is if the complexity is doubled, the hours per square foot may go up by 2.5 times.

USE OF EXISTING DATA

Some programs may use drawings and associated lists prepared on previous programs with no or few changes. If a significant (greater than 10 percent) usage of existing data is planned, then the total cost could be cut by 5 to 10 percent. Whether the lower or higher figure is used depends on how much confidence can be placed in the engineer's estimate. Also some costs will be incurred in retrieving, reviewing, checking, and processing these drawings. Therfore, some cost will be incurred even though the old drawing can be used exactly as is.

DRAWING REQUIREMENTS

If drawing quality, format and content requirements are more stringent than normal, costs should be adjusted higher. For example, the military calls out three forms of drawings: 1, 2, and 3. One company estimates the difference in cost for preparing each type of drawing as follows:

Form	Relative Cost	Remarks
3	Minimum base cost	Commercial, minimum quality
2	60 percent over base cost	Partial military requirements
1	25 percent higher than Form 2	Full military requirements to MIL-STD-100

Commercial companies may prepare drawings to their own standards, depending on the type of job; e.g., R&D, limited production, full-scale production, or factory test equipment. Each type of job would have different quality and content requirements for drawings tailored to the basic needs of the equipment being built. In this case, the company should have a standard multiplying factor to adjust the cost estimate to the correct value.

TEAM COMPETENCE

The types of engineers and designers used for the project make a difference in costs. If they are skilled and careful workers, drafting costs will be reduced because fewer drawing changes will be needed during the drawing life cycle. If on the other hand designers and engineers selected for the project are known for their many design changes and errors during a program, the total cost estimate should be raised to cover these additional expenses.

CASE IN POINT: A medium-size company received a contract to design and develop a new product for a customer. The project was assigned to an engineering team notorious for changing its drawings frequently before the design was frozen. The estimator tried to compensate for the additional drafting work by increasing the hours per drawing by 20% over normal estimates. At the end of the program the cost analysis indicated that the actual hours per drawing were 2½ times the normal estimate. The company consequently lost many thousands of dollars. Information like this, although after the fact, can help prevent repeating the same estimating mistake on the next job to be performed by this team, or least result in an investigation of why so much redesign was needed.

PROJECT SCHEDULE

A tight schedule means more pressure, less time for engineering and design, more overtime, and more changes and errors. Therefore, the cost estimate should be adjusted to account for inadequate or incomplete information to D&D that will require extra work to correct or follow-up on. Remember a rule in physics that it takes 4 times as much work or energy to move twice as fast. While not literally applicable to a project, this law applies at least partially to compressed projects.

DRAWING MIX

It is assumed that the drawing mix will be the same as for the historical data base. However, this may not be true; in some cases more mechanical detail parts or printed circuit boards, or more complicated assembly drawings may be needed. Therefore, the program should be evaluated in terms of the types and number of drawings to be prepared. If more simple drawings will be needed, the estimated cost can be revised downward.

DESIGN GROWTH

Depending on the project, a certain amount of additional drawings will be needed beyond the quantity specified by the engineer. The reason for this is unforeseen problems or obsolescence of drawings prepared during R&D but not usable during production. Depending on experience, hours should be increased by 10 to 20 percent for a typical project.

OUTSIDE SERVICES

If microfilming, aperture card preparation, photographic and drawing duplication work is done outside the company, these costs need to be added to the overall D&D estimate. (Assuming the costs aren't covered by another department.) These costs are included in the estimate as "other direct costs" (ODC), and are not subjected to overhead charges as direct labor costs are.

Other costs that should be included in ODC are polyester reproducibles, wash-off duplicates, computer time-sharing, computer-aided design services, specially formatted drawing vellums, PC board negatives, and other special materials.

RESEARCH TIME

Although this is also related to the project team described before, research time deserves special attention because D&D personnel may spend as much time looking for information as on the board. Therefore, the accessibility, accuracy, and timeliness of the information needed to prepare drawings must be assessed in terms of historical data.

D&D EFFICIENCY TREND

As described in Chapter 10, the efficiency* or productivity of the D&D department should be tracked so that the results of planning, control systems, and new techniques can be monitored and evaluated. If this is done, then the trend to higher or lower costs can be determined and future cost estimates adjusted to the projected costs for the next year. Of course, sufficient years of tracking data should exist to provide confidence that a trend is well established and not a sporadic occurrence connected with some other events not directly related to D&D operations.

CASE IN POINT: One company that kept adequate cost records was able to reduce its D&D estimates due to increasing efficiency of this operation. The results of a year of careful monitoring indicated that the average number of drafting hours per square foot of drawing produced had decreased from 11 to 8. The company was thus more competitive in its bids for new work

NATURE OF PRODUCT

A new product involving state-of-the-art technology requires more design and drafting labor than a product that requires no special or advanced techniques and components. Therefore, if the proposed project appears more complicated and uncertain in its implementation, additional hours should be added to the overall estimate.

9.5 Review of Estimate

Estimates should be reviewed with engineering and management to assure that D&D's interpretation of requirements is consistent with the needs of the program, manufacturing, and the customer. Specific areas that should be covered include:

1. Are all the drawings included in the estimate required?
2. Have needed drawings been omitted from the estimate?
3. Are the drawing quality and content excessive for needs and aims of project?

*Although some people claim that D&D operations are 75 percent efficient, the absolute efficiency is more likely about 50 percent.

4. Are basic drawing unit costs current or have they been invalidated due to changes in operating procedures, work scope, and costs?
5. Are costs covered in other departments duplicated in the D&D estimate?

Although little time is usually available for a thorough review of the estimate with management, it is a critical operation and can help avoid serious errors that could cause the company to lose money or a bid.

9.6 Cautionary Synopsis

No foolproof system for estimating exact D&D costs exists. However, by collecting accurate data from past projects, meaningful guides for estimating drafting costs on new projects can be formulated. These guides apply to the specific company involved and cannot be transferred to a new company without serious penalties. Application of historical cost data to new drawing efforts cannot be done blindly. Variable conditions and requirements must be assessed and the base estimate adjusted to compensate for these variables. In addition, historical cost data must be continually collected and reevaluated for a new cost basis. By doing this, the D&D manager can determine the effectiveness of his cost reduction program as a by product.

10

10.0 Management Control Systems, Schedules, and Reports

A brief review of some control techniques was given in Chapter 2. However, a more detailed discussion of other key areas for efficient management of the D&D department is needed.

Although the basic principles involved are essentially the same, each company has its own special methods and documents for controlling operations. Therefore, the material covered in this chapter, while haiving general applicability, will vary in its specific implementation from company to company.

BENEFITS AND APPLICABILITY OF THIS CHAPTER

This chapter reviews techniques for utilizing your manpower to its maximum and for keeping aware of department performance. Specific benefits include:

- Shows how a sample management control system cut drafting costs by about 30 percent.
- Describes various types of management reports.
- Presents a method for managing change in the D&D operation.
- Gives requirements for a management plan to help improve operations.

The techniques covered in this chapter are applicable to all size companies. However, smaller companies have to replace computer processing of data and reports with manual systems.

10.1 Management Control Systems For Improved Performance

A management control system measures and evaluates the use of people, money, materials, time, and space in achieving management goals. A critical part of this system is the supply of accurate information to management for evaluating performance periodically.

A prime feature of a control system is to provide timely, accurate and relevant data on progress and expenditures. These data are used by management to take effective measures to achieve objectives by correcting problems and improving operations. The quality of a management control system lies in its ability to promote efficiency, resolve conflicts, and control engineering data quality, quantity, and timeliness. Implementation of a management control system requires hard work and not taking anything for granted. These requirements are expressed by the Law of Entropy: An organization taken for granted inevitably deteriorates with time.[19]

CASE IN POINT: A manager of a department spent much of his time out of the plant, assuming that simple instructions and commands to his people would accomplish the prescribed work on time and within cost. Within a few months, schedules were slipping, costs were exceeding budgets, people were working overtime, and morale was low. Within 6 months, the company was losing over $100,000 and things were still in a chaotic mode.

Written clear-cut objectives, full-time follow-up of work, and rapid resolution of problems would have improved schedule and cost performance and could have resulted in a profit rather than a loss for the company. In addition, an effective management control system would have warned him of problems before they got too far out of hand.

MANUAL SYSTEMS

For many small and medium size companies, 100 percent manual management control systems are used because they don't need the speed, flexibility, and capacity of an automated data processing system. In addition, these companies usually can't afford the equipment and personnel for such a system. Consequently, traditional manual techniques are used consisting of:

1. Bar-type schedules to indicate tasks to be done and progress.
2. Periodic typewritten progress reports from responsible sources.

3. Project reviews with key staff members.
4. Cost reviews and controls by middle and top management.

The basic management control system is shown in Figure 43. It consists of a source of input data (perceiver), information shaping-manipulation and storage elements, evaluation and decision making elements, and feedback or corrective action elements.

Corrective action is taken when needed to eliminate ineffective operations. These actions include:

- Apply tighter and more frequent supervision and review of operations; i.e., more discipline.
- Replace ineffective staff members.
- Add more people to department (permanent/job shoppers).
- Provide new or larger facilities.
- Improve morale
- Install new equipment and materials.
- Revise operating procedures.
- Apply new techniques.
- Set up new communication networks.
- Improve skills and knowledge of staff.
- Institute new motivational elements.

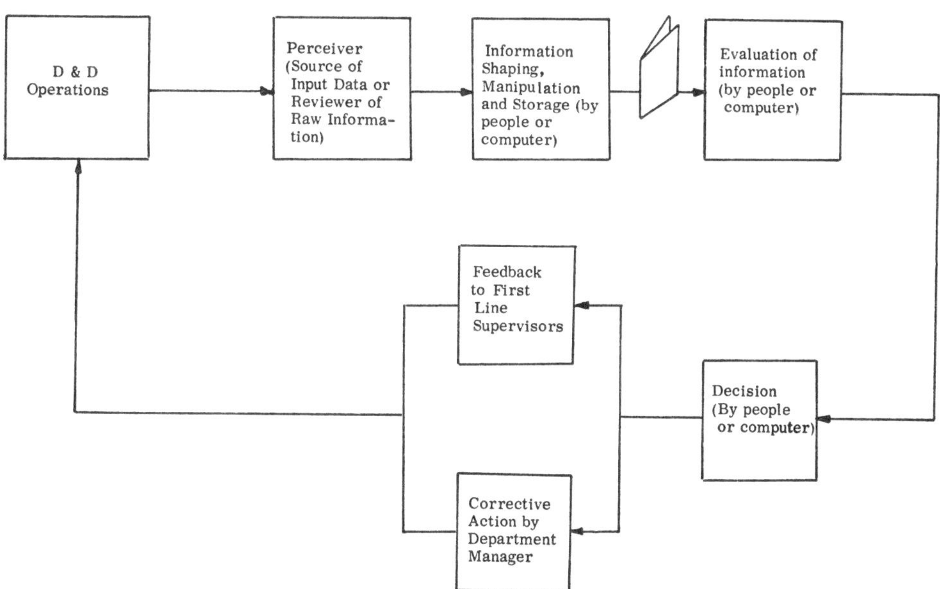

Figure 43. Management Control System

- Simplify operations.
- Add additional control elements.
- Formalize operational procedures.
- Send some work to outside D&D houses.

Since the basic steps for manual systems are the same as for automated systems, except for magnitude of labor, quantity of data processed, and equipment used, further discussion of manual operations is omitted here in preference to proceeding to an automated or computerized management control system. However, manual scheduling and reporting techniques are discussed in subsequent sections of this chapter. Other manual control methods were described in Chapter 2.

COMPUTER SYSTEMS

When the term "computer system" is used, keep in mind that the whole management control system is not computerized. A more correct term is computer-assisted management control system because humans must still input the raw data, make the final evaluation and decisions, and implement required corrective actions. Thus, most activites described for manual systems also apply to computer management systems.

ADVANTAGES AND DISADVANTAGES—The advantages of computer-assisted management control systems are speed, flexibility in processing information, large capacity, and lower cost when the amount of information processing is considerable. In addition, outputs not specifically required are often a by-product; e.g., with some additional programming, special reports can be created from the basic data base. Although not currently used widely, computers can also provide decision-making capabilities.

The disadvantages of computer systems are that they take detailed analysis of requirements, extra planning, and effective implementation to assure a cost-and performance-effective system. The technical expertise is frequently not available to accomplish these tasks well; consequently, the results often fail to meet management expectations. Other problems include delays and excessive costs not forecast in the original decision to move into automated systems. In addition, if a new system is being implemented, staff resistance, if not handled properly, can undermine an otherwise well-designed system and result in its failure.

Many computer systems in the US have failed to live up to the expectations of their users.[20] In some cases, companies have reverted to manual systems because of improperly implemented computer operations or have taken several years to recover from an inept introduction of the computer system into company operations.

EXAMPLE OF A COMPUTER-ASSISTED SYSTEM—An example of a computer-assisted management control system is described next.

> **CASE IN POINT: One company instituted a computer-assisted management control system that reduced drafting costs by 30 percent in a two-year period. The system employed weekly efficiency reports and short-interval scheduling to mobilize its staff into a hard working, well disciplined group.**

SYSTEM DESCRIPTION: Figure 44 shows a computer-assisted control system. The input to the system includes change-in-status reports and new data inputted to D&D. This information is written on status cards and filed in an open file for daily reference and updating. Information for only one document is recorded on a card. The card contains document number and revision letter, estimate to complete work for each operation, start and stop days and hours, job number and time to actually complete the work. These cards are revised whenever the document status changes and thus provide current information on critical documents of concern to engineering and management.

At the end of the week, these handwritten cards are submitted to the data processing facility for conversion to keypunched and interpreted computer cards. Keypunched cards are then checked and inputted into the computer, which is programmed to process the data and provide various types of reports. These reports include:

- Listing of all documents in D&D and their status, job number, estimated and actual hours to complete.
- Listing of all documents by project or work order.
- Listing of all change documents (EO/CN's) in process against a particular drawing or parts list.
- Efficiency reports.
- Department output in terms of number of drawing, PL, EO, and test specification sheets.

These reports are then distributed to key management personnel for evaluation and action if needed. When problems or unusual developments occur, the D&D manager takes action by discussing the problem with his supervisors and deciding on the best course of action to rectify the undesirable condition.

ANALYSIS AND DOCUMENTATION OF WORK UNITS: Before implementing this system, analysis of operations is performed to determine standard work units (in hours) for each type of document prepared. A manual is prepared based on this analysis for providing estimators with data for determining the hours required to prepare a drawing, engineering order, and printed

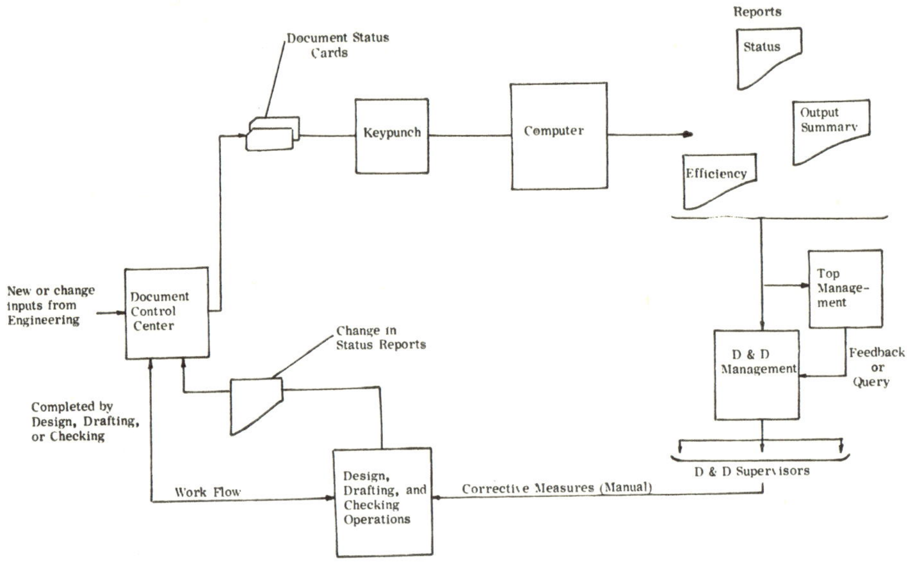

Figure 44. Computer-Assisted Control System

circuit artwork. Hours to revise or redraw a document are also designated. Thus, as work is received by the D&D department, the hours to prepare the document are obtained from the manual and recorded on the work package and status card.

Besides giving the staff guidelines for completing their work, the estimates are compiled by the computer to indicate the weekly work load inputted to the D&D department.

RECORDING DOCUMENT STATUS: When the work package is assigned by the supervisor to a designer, draftsman, or checker, the person receiving the work is identified on the status card, along with the time of issue and other control/administrative information. When the designer completes his assignment, he returns the package to the next functional supervisor and submits a change in status report to the clerk who maintains the status cards. The clerk adds the completion time, the actual hours charged, and the next function that is to receive the work package such as drafting. When the draftsman receives the document, he submits a change in status report and the clerk updates the status card. Thus, every step of the process is tracked and recorded, providing a complete summary of the documents's history, location, hours charged, and who worked on it. If unusual requirements or changes in input from engineering occur, these are also recorded on the card and estimates revised to account for the additional work.

Of course, the above inputting of data can be done using a terminal that inputs status information directly into a computer data bank located in the data processing facility. Thus conventional records are unnecessary and information on various documents can be displayed on a CRT, which is part of the terminal, when requested.

REPORTS: Several reports can be obtained by management to review and monitor D&D performance. These include efficiency, work load by job, and output by function.

An efficiency report by employee number gives the number of drawings and sheets completed, estimated hours assigned versus actual hours charged, and hours charged to document but not included in estimate (target). The ratio of actual to estimated hours gives employee efficiency, which can exceed 100 percent using this system. (The estimated hours are based on an average experienced person performing the task; therefore, above average people will exceed 100 percent efficiency while new and inexperienced people will achieve less than 100 percent.)

Figure 45 shows a work load summary for the drafting function. It includes the cost center, basic documents, EO/CN's, estimated hours, employee number, start and stop dates, elapsed days, actual hours, and efficiency for each document. This summary also indicates when a document has not been assigned to D&D personnel (work not started).

Other reports include the outputs of each function within the D&D department: design, drafting, and checking. Thus, they give the number of drawings completed by design, drafting, and check, including change documents, for all jobs in the department.

USE OF REPORTS: Once the reports have been received by management, evaluation and responses to their contents are performed manually. To detect problems, however, takes experience with the operation so that unusual trends or occurrences can be identified by the manager and analyzed for their impact on the project or department. Thus, minor or transient changes can be identified for what they are and ignored. Tracking down every minor perturbation is impractical and time consuming. Therefore, the manager must be able to separate significant trends from minor ones.

CORRECTIVE ACTIONS: Corrective actions are the same as described for manual systems.

10.2 Scheduling Techniques

Although scheduling techniques are an integral part of management control systems, they are presented here separately to allow a more independent analysis.

The purpose of scheduling is to provide a map of what and when different tasks will be completed so that the job or project will be completed on time. Its

DRAFTING WORK-LOAD SUMMARY

PAGE 5
WEEK ENDING- 26 AUGUST 197

CC CENT	DRAWING NUMBER	SIZE	SHEETS	EO/DCN NUMBER	G.O. NUMBER	ESTIMATED HOURS	EMPLOYEE NUMBER	START DATE	START TIME	STOP DATE	STOP TIME	ELAPSED DAYS	ACTUAL HOURS	% EFF
11	8100001686	J	1		P5563	1.4	* NOT ASSIGNED							
11	100001717	J	1		G392	4.6	14091	8-20-	300	8-21-	1000	2	5.0	92.0
11	100001727	J	1		G420	2.5	* NOT ASSIGNED							
11	100001736	J	1		G420	15.4	14075	8-17-	100	8-22-	430	6	32.8	47.0
11	100001750	D	1		G420	2.4	4	8-21-	200	8-21-	430	1	2.5	96.0
11	100001751	F	1		E213	6.3	13948	9-20-	215	8-20-	500	1	10.7	58.9
11	100001752	D	1		G420	2.4	3	8-21-	130	8-21-	345	1	2.3	104.3
11	110000629	F	1		G320	3.0	* NOT ASSIGNED							
11	150000466	D	1		G420	2.3	14091	9-17-	800	9-17-	300	1	6.5	35.4
11	150000468	D	1		G420	2.2	14075	8-15-	430	9-16-	130	2	4.5	48.9
11	8160000403	J	1		G385	1.3	* NOT ASSIGNED							
11	190000711	A	6		F209	2.9	14077	8-22-	1100	8-23-	500	2	0.5	580.0
11	190000718	A	7		G420	1.0	13948	8-14-	915	8-14-	1200	1	2.7	37.0
11	190000723	A	10		G420	8.0	* NOT ASSIGNED							
11	190000724	A	10		G420	3.0	13971	8-23-	400	* NOT COMPLETED				
11	190000738	A	21		G420	6.0	3	8-23-	1030	8-24-	1030	2	7.0	85.7
11	190000744	A	19		G420	8.0	14077	9-17-	1130	8-21-	400	5	19.0	42.1
11	879000037	A	0		209	0.0	* NOT ASSIGNED							
11	FILE		0		G339	8.5	13910	8-16-	430	8-16-	500	1	8.5	100.0
11	MP-FX BD1PWRSUP	C	1		G368	2.2	* NOT ASSIGNED							
11	MPSKS-19593	D	1		G383	3.7	* NOT ASSIGNED							
11	MPSKS19813	C	1		F293	1.9	* NOT ASSIGNED							
11	MPSKS19814	C	1		F293	1.6	* NOT ASSIGNED							
11	MPSKS19819	C	1		F293	2.2	13910	6-26-	410	* NOT COMPLETED				
11	MPSKS19830	C	1		F293	7.6	14142	8-21-	312	* NOT COMPLETED				
11	MPSKS23331-1	C	1		F201	3.1	14337	9-23-	130	8-23-	500	1	3.5	88.6
11	MPSKS23642	D	1		F201	6.7	* NOT ASSIGNED							
11	MPSKS23670	D	1		F201	7.3	14337	8-15-	1030	8-16-	230	2	11.0	66.4
11	MPSKS23710	D	1		F201	11.9	14142	8- 2-	210	* NOT COMPLETED				
11	MPSKS23745		0		F201	10.6	13910	9-15-	1040	9-16-	215	2	10.6	100.0
11	MPSWITCH		0		F293	5.0	* NOT ASSIGNED							
11	MP19680	F	1		F201	9.6	* NOT ASSIGNED							
11	MP2030001767	C	1		E209	4.3	* NOT ASSIGNED							
11	MP8080002349	D	1		G420	3.7	14337	8-17-	230	8-17-	500	1	5.5	67.3
11	MP8080002350	F	2		G313	0.5	* NOT ASSIGNED							
11	MP8080002364	C	1		G420	3.1	14337	8-23-	900	8-23-	130	1	3.5	88.6
11	MP8080002367	D	1		E213	11.9	14337	8-23-	930	8-23-	900	1	13.0	91.5
11	MP8080002368	D	1		E213	13.7	* NOT ASSIGNED							
11	MP8080002369		0		F213	11.6	7681	8-17-	300	8-20-	400	4	10.0	116.0
11	MP8080002370		0		E213	13.1	7681	8-20-	400	8-24-	900	5	20.5	63.9
11	MP8080002371	D	2		F213	13.7	* NOT ASSIGNED							
11	MP8080002377	F	1		F213	7.6	14142	8-20-	242	9-21-	1036	2	4.9	155.1
11	MP8080002379	F	2		F213	14.9	* NOT ASSIGNED							
11	MP8080002380	D	2		F213	13.7	14142	8-21-	312	* NOT COMPLETED				
11	PLDUMMY	B	101		G313	10.0	* NOT ASSIGNED							

Figure 45. Work Load Summary

prime value lies in its ability to provide a simple visual means for management to monitor progress and to flag specific delays before they become so serious that the overall effort is bogged down.

Requirements for an effective scheduling technique are the same as described for management control systems, for example, simple, inexpensive, effective, and easy to maintain. The scheduling techniques described next are the bar and tracking charts.

BAR CHART

A bar chart is shown in Figure 46. Many variations exist but they all basically plot tasks to be completed in a vertical column and show start and stop times along horizontal rows for each task. Figure 46 shows the overall drawing effort for the D&D department. As time passes, the hollow bar is filled in to show the amount of progress. In addition, the upper area showing the months is filled in to give the total elapsed time for the project. If the blackened area for the drawing task is behind the elapsed months, then the work is behind schedule. If it is ahead of June, then it is ahead of schedule. This type of schedule is very rough and only shows overall progress. If detailed progress is needed on specific subtasks, then a more detailed chart identifying each drawing is used.

TRACKING CHART

A tracking chart is a first line supervisor's work and schedule sheet. He can keep track of all hot documents conveniently on a few A-size sheets that are easy to handle, maintain, and copy. Each drawing is listed with its due date in design/drafting, location, status, and estimated release date. A remarks column is used to identify new engineering inputs, problems, or changes in status. Because it may contain more than 100 documents, a tracking chart is not a convenient method for reporting progress to management. However, it can be an extremely effective control and scheduling tool.

> **CASE IN POINT: One medium-size company used tracking charts effectively and managed to complete an extremely important job on schedule using this technique. The chart contained over 100 drawings and associated data that had to be prepared in order for manufacturing to meet its delivery schedule.**
>
> **The drafting supervisor prepared this sheet at the start of the program, determined how many documents had to be prepared daily to meet the desired end date, and updated it every day. Although it took a lot of his time to maintain this**

SCHEDULE OF EVENTS	1974												1975								
PROJECT / MONTH	J	F	M	A	M	J	J	A	S	O	N	D	J	F	M	A	M	J	J	A	S
PROJECT AR-88																					
PROJECT B-42																					
PROJECT AC-100																					
PROJECT BLU-74																					
PROJECT 200-WL																					
PROJECT S-30B																					
PROJECT F-200																					

Figure 46. Bar Chart

tracking chart, it kept the group mobilized and with firm objectives in mind. As a result, all documents were released 2 weeks ahead of schedule.

10.3 Special Reports for Better Planning

MANPOWER REQUIREMENTS SUMMARY

A manpower requirements summary identifies all jobs/projects in house and the number of people required on a monthly basis to support the job until it is completed. The sum of the manpower needed for all jobs gives the total staff required in the D&D department for each month.

This summary allows the manager to anticipate fluctuations in manpower needs over the next 6 to 12 months. Thus, he can make special efforts to obtain additional work when the presently known work requirements are inadequate to maintain his present staff. Or the opposite may be true; i.e., the work load may be greater than his staff and he must either hire more people or get job shoppers.

MONTHLY BUDGET REPORT

A monthly budget report identifies the various expenditures for the department. It gives the following information:

- Budgeted dollars for various expenditure elements for the preceding month.
- Actual dollars expended for that month.
- Variance between budgeted and actual dollars.

Budget elements included in the report are direct and indirect labor, supplies, materials, telephone, fringe benefits, recruiting, and travel expenses.

In addition to the monthly overhead rate, a year-to-date summary of all budgeted and actual costs/expenditures is given.

ANTICIPATED FINAL COST REPORT

The anticipated final cost report is prepared primarily for higher management so that it can evaluate and budget for various jobs in progress. This report gives each manager's estimate of what it will take to complete his work or project. It thus flags overrun conditions and identifies the estimated magnitude of the overrun. The report may include the following information:

1. Cost elements (drafting, design, checking, and standards)
2. Cumulative actual expenditures during previous years
3. Cumulative actual expenditures through last year
4. Actual costs for this year on monthly basis from January through some period
5. Forecasted cost for remainder of year
6. Balance of cost to complete beyond monthly forecast period
7. Total estimated cost to complete
8. Current anticipated cost to complete (sum of cumulatives for preceding years, last year, and this year)
9. The original cost budgeted for the job.

The anticipated final cost report is usually made up for a whole project and therefore includes design and drafting cost data.

Besides helping management adjust plans to accommodate over or under run conditions, this report keeps the program manager and engineer constantly aware of their performance and provides information for making better estimates in the future or finding a way to cut costs in order to stay within budget.

10.4 Change Management

Managing change is an everpresent and critical function in modern industry. Managers are constantly being flooded with information on new equipment, techniques, and materials that are reputed to reduce costs or improve performance. To avoid costly mistakes, systematic evaluation, approval, and implementation of a proposed change is needed.

The key to effective change management is a centralized and formalized control system that includes the correct levels of technical and management personnel. In addition, a good cross-section of company operations is needed to ensure objective and company-wide evaluation of the effectiveness and impact of the change on related areas.

In evaluating the value of a new system, or procedures, refer to Chapter 1 for guidelines. Remember that the method of handling people during changes is of major importance to avoid a non-cooperative attitude or loss of good people. Failure to adequately prepare people for changes is perhaps one of the major blocks to successful change management and implementation.

CASE IN POINT: One company, under pressure to reduce cost, jumped into introducing several major changes in the operation of the D&D group without proper coordination with its key people. As a result, the staff became apprehensive about their jobs and future and began leaving the company. Within 6 months, several experienced people left the company, seriously reducing the capability of the department and undermining the effectiveness of the new systems.

This situation could have been made less likely by adequate communication to the staff at all levels of what changes were to be made, how, when, and why. In addition, explicit reassurance that no one would be hurt by the changes was needed. Of course, statements that no one would be hurt by the changes would not be effective unless the staff trusted its management.

A system change request form for controlling internal changes to company operations is a necessary item. The form forces the change originator to think through the various aspects of his idea. It thus avoids wasting other's time by doing the thinking for him and avoids overlooking important factors that should be considered before the change is submitted to higher management.

Evaluation and approval of a major change should include review by a change control board (CCB) consisting of key department and company people. This allows various opinions to be presented on the value of the change and avoids overlooking serious interface problems later on downstream.

Another advantage of having a CCB is that is offers the opportunity for CCB members to tell about their experience with similar systems or procedures at other companies and what problems were incurred.

Once approved, a change should be reviewed periodically by the CCB to assure that the new process, equipment, or material is being implemented as originally approved by the board. This allows the CCB to make sure it gets what was desired or to find out why the change hasn't been implemented as originally agreed to. It also helps it avoid making similar mistakes in the future.

10.5 Management Plan

A management plan provides the department with its charter of objectives, philosophy, organization, and techniques of operation. Besides providing a previously agreed to baseline for determining department performance and problem solutions, it avoids confusion, working at cross-purposes, and focusing on the wrong things. It helps the manager crystallize his ideas about the coming year and what he feels are the most important things that should be done. It also helps him communicate these ideas to his staff with a minimum of ambiguity and changes because he forgot what they were 6 months ago.

This plan is also a good method for improving communication among key staff members and providing them with an opportunity to challenge unreasonable objectives, solutions to problems, and organizational assignments.

Another benefit from a management plan is that it can help a new member of the department to learn about its operations and key goals.

The topics covered in a management plan depend on what the manager feels is important. However, some possible subjects are listed next:

1. Department objectives
2 Management philosophy
3. Organization and staff
4. Control systems
5. Type of work performed and scope
6. Key functions of various staff members
7. Interface requirements with other departments
8. Methods of meeting objectives
9. Problem areas requiring resolution
10. Last year's achievements and failures
11. Schedule for meeting objectives
12. Change management requirements
13. Contract/job analysis procedure
14. Miscellaneous topics

Although many pages can be written covering the preceding topics, the management plan should not be allowed to grow too large. It should highlight key topics and reference the reader to more detailed procedures in the DRM or other company procedures manual. Too long a plan will discourage its use and may not be read carefully by the staff. To assure a minimum exposure to the plan, a checkprint should be issued to the staff and a meeting held after they have had time to study it. Each major point presented in the plan should be discussed and the department manager should encourage differing viewpoints to be brought out. After revision, the final management plan should be issued to the staff.

Note that the management plan is a dynamic tool for achieving objectives. Therefore, it should be updated whenever serious changes occur. Of course, a new plan should be issued at least each year, even if much of its content is unchanged because most staff members will have forgotten much of its content in 12 months.

REFERENCES

[1]George E. Rowbotham, "Drafting Management," *Graphic Science*, May 1969, p. 8-12

[2]"Centralized vs Team Drafting," *Reprographics*, June 1972, p. 8

[3]William H. Rideout, "Achieving Better Utilization of Engineering Manpower," 7th Annual Engineering Graphics Management Seminar at University of Southern California, Jan. 30, 1970

[4]Raymond J. Behan, "Short Interval Scheduling Cuts Design Time," *The Successful Engineer-Manager* (Hayden, 1971), p. 100-104

[5]L.B. Linksy, "Automated Drafting System Study and Implementation," Raytheon Report, 1973

[6]"From Central Engineering Data Bank," *Reprographics*, October 1971, p. 10

[7]Jasper U. Teague, "Increase Profit with Improved Engineering Standards," 8th Annual Engineering Graphics Management Seminar at University of Southern California, January, 1971

[8]Chester F. Stiles, "Cost Reduction: A Continuing Project," *Reproduction Methods*, January 1969, p. 46

[9]"Microfilming Shrinks Highway Department Costs," *Engineering Graphics*, July 1971, p. 15

[10]"Efficient Engineering Files Cut Costs For Manufacturing," *Reprograhics*, March 1969, p. 18-19

[11]Ron Biggie, "GTE Sylvania Utilizes Microfilm to Meet Paperwork Requirements," *Reprographics*, March 1972, p. 23-27

[12]"Detail Die Drawings," *Plan and Print*, July 1971, Vol. 44, No. 7

[13]"Erasing the Retrace Drafting Chores at Borg Warner," *Engineering Graphics*, March 1973, p. 12-15

[14]J.A. Buffa, "Typing on Engineering Documents," *Graphic Science*, June 1971, p. 18-19

[15]Bruce Grant & Harold Goody, "Why Draft With Ink on Film?" *Graphic Science*, February 1970, p. 22-24

[16]H. Gilson, "Small Plant Engineering Records Provide Evidence for Patents," *Technical Aids for Small Manufacturers*, No. 88, October 1964

[17]Ron Biggie, "GTE Sylvania Utilizes Mirofilm to Meet Paperwork Requirements," *Reprographics*, March 1972, p. 23-27

[18]"Microfilm System Saves Time and Money," *Graphic Science*, January 1971, p. 10-12

[19]T. Samaras, "Measuring Organizational Entropy," *Akron Business and Economic Review*, Summer of 1973, p. 15-21

[20]L.K. Albrecht, *Organization and Management of Information Processing Systems* (New York, Macmillan: 1973), p. v

APPENDIX

Other organizations in engineering graphics and related fields include:*

TECHNICAL GRAPHICS

American Engineering Model Society (AEMS)
P.O. Box 177
Ross, Ohio 45061

Design and Drafting Management Council
P.O. Box 11811
Santa Ana, Ca. 92711

American Congress on Surveying & Mapping (ACSM)
430 Woodward Bldg.
733 15th St. N.W.
Wash. D.C. 20006

The Institute of Draftsmen, Australia
36 Filbert St.
South Caulfield, 3162
Victoria, Australia

Peninsula Drafting Management Association Inc.
P.O. Box 11271
Palo Alto, Ca. 94306

Industrial Graphics International (IGI)
P.O. Box 4046
Huntsville, Ala. 35802

National Microfilm Association (NMA)
8728 Coleville Rd.
Silver Springs, Md. 20910

Association of Reproduction Materials Manufacturers Inc.
901 N. Washington St.
Alexandria, Va. 22313

International Association of Graphic Communications
P.O. Box 84
Los Alamitos, Ca. 90720

TECHNICAL DOCUMENTATION

Society for Technical Communication, Inc. (STC)
1010 Vermont Ave. N.W.
Los Angeles, Ca. 90000

American Society for Engineering Education (ASEE)
2100 Pennsylvania Ave. N.W.
Washington, D.C. 20037

National Technical Services Association (NTSA)
1225 Connecticut Ave. N.W.
Washington, D.C. 20036

The Metric Association
2004 Ash St.
Waukegan, Ill. 60085

Standards Engineers Society
P.O. Box 7505
Philadelphia, Pa. 19101

*Adapted from *Engineering Graphics*, 1975.

American National Standards
Institute (ANSI)
1430 Broadway
New York, N.Y. 10018

American Federation of
Information Processing
Societies (AFIPS)
210 Summit Avenue
Montvale, N.J. 07645

Institute for Graphic
Communication, Inc.
520 Commonwealth Ave.
Boston, Mass. 02215

American Ordnance Assn.
(AOA)
Engineering Documentation
Section Computer-Aided
Design Section
Transportation Bldg.
Washington, D.C. 20006

INDEX

A

B

C

D

E

J

K

L

M

N

O

P

Q

R

S

T

U

V

W

X